Genetics of Epilepsy and Refractory Epilepsy

Colloquium
Digital Library of Life Sciences

This e-book is a copyrighted work in the Colloquium Digital Library—an innovative collection of time saving references and tools for researchers and students who want to quickly get up to speed in a new area or fundamental biomedical/life sciences topic. Each PDF e-book in the collection is an in-depth overview of a fast-moving or fundamental area of research, authored by a prominent contributor to the field. We call these e-books *Lectures* because they are intended for a broad, diverse audience of life scientists, in the spirit of a plenary lecture delivered by a keynote speaker or visiting professor. Individual e-books are published as contributions to a particular thematic **series**, each covering a different subject area and managed by its own prestigious editor, who oversees topic and author selection as well as scientific review. Readers are invited to see highlights of fields other than their own, keep up with advances in various disciplines, and refresh their understanding of core concepts in cell & molecular biology.

For the full list of published and forthcoming Lectures, please visit the Colloquium homepage: www.morganclaypool.com/page/lifesci

Access to the Colloquium Digital Library is available by institutional license. Please e-mail info@morganclaypool.com for more information.

Morgan & Claypool Life Sciences is a signatory to the STM Permission Guidelines. All figures used with permission.

Colloquium Series on
The Genetic Basis of Human Disease

Editor

Michael Dean, Ph.D., *Head, Human Genetics Section, Senior Investigator, Laboratory of Experimental Immunology, National Cancer Institute (at Frederick)*

This series will explore the genetic basis of human disease, documenting the molecular basis for rare, common, Mendelian, and complex conditions. The series will overview the fundamental principles in understanding concepts such as Mendel's laws of inheritance and genetic mapping through modern examples. In addition, current methods (GWAS, genome sequencing) and hot topics (epigenetics, imprinting) will be introduced through examples of specific diseases.

For a full list of published and forthcoming titles:
http://www.morganclaypool.com/page/gbhd

Genetics of Epilepsy and Refractory Epilepsy
Alberto Lazarowski and Liliana Czornyj
www.morganclaypool.com

ISBN: 9781615045327 paperback

ISBN: 9781615045334 ebook

DOI: 10.4199/C00073ED1V01Y201303GBD002

A Publication in the

COLLOQUIUM SERIES ON THE GENETIC BASIS OF HUMAN DISEASE

Lecture #2

Series Editor: Michael Dean, National Cancer Institute

Series ISSN:
ISSN 2168-4006   print
ISSN 2168-4022   electronic

Genetics of Epilepsy and Refractory Epilepsy

Alberto Lazarowski
Universidad de Buenos Aires
Instituto de Fisiopatología y Bioquímica Clínica - INFIBIOC
Instituto de Biología Celular y Neurociencias "Prof. E. de Robertis"
Fundación Investigar

Liliana Czornyj
Hospital de Pediatría S.A.M.I.C. "Prof. Dr. Juan P. Garrahan"

COLLOQUIUM SERIES ON THE GENETIC BASIS OF HUMAN DISEASE #2

MORGAN & CLAYPOOL LIFE SCIENCES

ABSTRACT

Epilepsy affects approximately 3% of the population, and is usually defined as a tendency to experience recurrent seizures arising from periodic neuronal hyperexcitability of unknown causes. Different genetic factors, through various mechanisms, can cause this abnormal neuronal behavior. The etiology of epilepsy is a major determinant of clinical course and prognosis. Many of the genes that have been implicated in idiopathic epilepsies code for ion channels, whereas a wide spectrum of syndromes where epilepsy is a main clinical feature are caused by mutated genes that are involved in functions as diverse as cortical development, brain malformations, mitochondrial function, and cell metabolism. Similarly, different conditions as hypoxia, trauma, infections, or metabolic unbalances can develop epileptic syndromes where upregulation of several genes could be related to the epileptogenic mechanisms. The most common human genetic epilepsies display a complex pattern of inheritance, and the susceptible genes are largely unknown. However, major advances have recently been made in our understanding of the genetic basis of monogenic inherited epilepsies. As we continue to unravel the molecular genetic basis for epilepsies, it will increasingly influence their classification and diagnosis. A majority of epileptic patients may control their crisis with anticonvulsant drugs, however 30%–40% became refractory to pharmacological therapies and require surgical treatment. The challenge of the molecular revolution will be the design of the best treatment protocols based on genetic profiles that include both the specific mechanistic etiology of the epilepsies, as well as their potential refractory behavior to current medications. This includes also the design of new therapeutic agents and targets, so as to reduce the number of cases with refractory epilepsy and epileptogenesis, and perhaps avoid the current surgical treatment (a procedure that was first described more than 4000 years ago) except as a last option.

KEYWORDS

epilepsy, genetics, mutations on ion channels, GABA receptor, glutamate receptors, acetylcholin receptors, glycine receptors, metabolic diseases, acquired epilepsies, pharmacokinetics, pharmacodynamics, pharmacogenetics, drug transporters, antiepileptic drugs (AEDs), AED resistance, refractory epilepsy, multidrug resistant genes, epileptogenesis

Contents

Contents

CHAPTER 1

Introduction

Only the right dose differentiates between a poison and a remedy
—Paracelsus, 1493–1541

Epilepsy is a common neurological condition that has been in existence for ages and continues to affect approximately 50 million individuals worldwide [1, 2], involving recurring and unprovoked seizures. Epileptic seizures have been defined as abnormal, excessive, and synchronous neuronal activity in the brain [3]. To understand the genetics of epilepsies, a clear description of the clinical terminology and its classification is required. According to the International League Against Epilepsy (ILAE), epilepsies are classified as idiopathic, symptomatic, or cryptogenic, and all of them can be either generalized or focal. Although the legacy of epilepsy was hypothesized a long time ago, currently, it is accepted that genetic factors play a central role in the idiopathic generalized epilepsies (IGEs).

The seizures themselves are the clinical manifestation of an underlying transient abnormality of cortical neuronal activity. The phenotypic expression of each seizure is determined by the point of origin of the hyperexcitability and its degree of spread in the brain. Thus, seizure effects can vary from loss of awareness to more obvious and distressing tonic–clonic manifestations.

The seizures, which can last between a few seconds and a few minutes, can be isolated or can occur in series. The causes of sporadic or recurrent seizures are numerous, and they include several etiologies such as acquired structural brain damage, altered metabolic states, or inborn brain malformations; however, around 1% of all people develop epilepsy (recurrent unprovoked seizures) for no obvious reason and without any other neurological abnormalities. Both acquired and idiopathic forms present genetic differences as compared with nonconvulsive individuals.

In some cases, the genetic point of origin has already been shown by successful cloning of the mutated gene. The first gene that was found to be responsible for idiopathic epilepsy in humans is the neuronal nicotinic acetylcholine receptor (nAChR) α4 subunit (*CHRNA4*), (α4β2 hetero-oligomer), which is associated with autosomal dominant nocturnal frontal lobe epilepsy (ADNFLE) (Figure 1). A missense mutation in the gene leads to high affinity of ligand binding and slow desensitization [4].

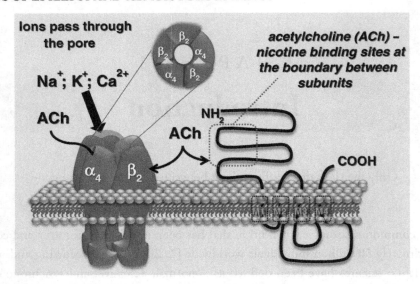

FIGURE 1: Schematic representation of nAChRs. The nAChR is a ligand-gated ion channel and is composed of five (α or β) subunits that can be homomeric (all α or all β) or heteromeric (mixture of α and β). Each subunit consists of a large amino-terminal extracellular domain with a transmembrane domain and a variable cytoplasmic domain that include a long cytoplasmic loop between M3 and M4 and other shorter loops connecting the domains. Sites for allosteric modulators are located in four hydrophobic transmembrane domains the transmembrane domains (M1–M4). CHRNA4 and CHRNB2 mutations produce gain-of-function (GOF) effects. The α4 subunit of the nAChR is encoded by the CHRNA4 gene, which is one of the genes that are mutated in ADNFLE.

A wide spectrum of syndromes—as diverse as neurodegenerative disorders, mental retardation syndromes, as well as neuronal migration disorders and mitochondrial encephalomyopathies—have been described as capable of developing severe epileptic phenotypes. Moreover, in addition to the large group of idiopathic epilepsies, more than 200 single-gene disorders are known in which the presence of recurrent seizures are an important part of the phenotype. One intriguing feature is that all possible modes of inheritance, including autosomal, X-chromosomal, mitochondrial, and complex inheritance, have been described in epilepsies. Furthermore, under a strict model of autosomal dominance, the mutant alleles of these genes should be expected to cause epilepsy in each of their carriers; however, family studies show that the penetrance of these mutations, the age of onset, as well as the severity of the phenotype, varies within families [5, 6].

A genetic contribution to etiology has been estimated to be present in about 40% of patients with epilepsy. The genetic control of neuronal synchrony may be direct or indirect, and the various approaches to the classification of genetic epilepsies reflect this.

It is useful to categorize genetic epilepsies according to the mechanisms of inheritance involved. This identifies three major groups:

- *Mendelian disorders*, in which a single major locus can account for segregation of the disease trait.
- *Non-Mendelian or "complex" diseases*, in which the pattern of familial clustering can be accounted for by the interaction of several loci together with environmental factors, or by the maternal inheritance pattern of mitochondrial DNA (mtDNA).
- *Chromosomal disorders*, in which a gross cytogenetic abnormality is present.

A second useful distinction is between the "symptomatic" epilepsies, in which recurrent seizures are one component of a multifaceted neurological phenotype, and the "idiopathic" epilepsies, in which recurrent seizures occur in individuals who are otherwise neurologically and cognitively intact and who have no detectable anatomical or metabolic abnormality.

Although research over the years has led to significant advances in understanding the pathophysiology of epilepsy, the specific causes of several types of epilepsy are unknown [7].

More recently, Simon D. Shovron critically discusses the etiological classification of different epileptic syndromes defined as

(a) *Idiopathic epilepsy:* epilepsy of predominately genetic or presumed genetic origin.
(b) *Symptomatic epilepsy:* epilepsy of an acquired or genetic cause, associated with gross anatomical or pathological abnormalities, and/or clinical features, indicative of underlying disease or condition.
(c) *Provoked epilepsy:* epilepsy in which a specific systemic or environmental factor is the predominant cause of the seizures and in which there are no gross causative neuroanatomical or neuropathological changes.
(d) *Cryptogenic epilepsy:* epilepsy of presumed symptomatic nature in which the cause has not been identified.

It is clear that there are cases for which applying this categorization is difficult and that the results are sometimes arbitrary (Figure 2). Interestingly, some of the childhood syndromes are classified as idiopathic epilepsies, without any verified genetic basis, as for instance, the benign rolandic

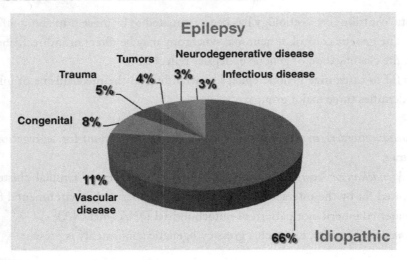

FIGURE 2: Frequency of different etiologies for acquired epilepsies.

epilepsy. Meanwhile, for the Lennox–Gastaut syndrome or several forms of West syndromes there are presumptions of a genetic cause, however, they are included in the category of symptomatic epilepsy [8]. Furthermore, the ILAE has suggested that all these etiological categories of the epilepsy: idiopathic, symptomatic, and cryptogenic should be replaced by new denominations: genetic, structural/metabolic, and unknown, respectively, but some authors have expressed their disagreement [8–11], and this chapter of the definitions of epilepsy etiology still remains to be finalized.

Although antiepileptic drugs (AEDs) are the primary option for the management of epilepsy, several clinical studies have shown that high seizure frequency before the onset of therapy is associated with an increased disease severity and predicts a poor outcome of pharmacotherapy. These patients also tend to develop a refractory epilepsy (RE), that may be defined as failure of adequate response to two tolerated and appropriately chosen and used AED schedules (whether as monotherapy or in combination). The "intrinsic severity hypothesis" implies that (still unknown) neurobiological factors that contribute to increased disease severity may also play a role in pharmacoresistance [12]. Furthermore, categorization of cause according to mechanism should be at the core of any classification scheme [8], focusing on the suspected close relationship between epileptogenesis, intrinsic severity and refractoriness. Moreover, the disease is often accompanied by neurobiologic, cognitive, psychological, and behavioral changes that *per se* may heighten susceptibility to more frequent and severe seizures developing simultaneously with drug resistance in epilepsy.

In this context, the molecular and biological mechanisms of disturbances in several neurological processes could be the basis for the increased risk to repetitive expression of seizures as well as for the mechanisms that potentially underlies drug resistance in epilepsy.

What Is Epilepsy?

One in 10 persons with a normal life span can expect to experience at least one seizure.

Fisher et al. defined epileptic seizures as abnormal, excessive, and synchronous neuronal activity in the brain that manifest themselves in different ways depending on their site of origin and subsequent spread. The clinical diversity observed in epileptic seizure disorders is a reflection of the numerous cellular and network routes to seizure genesis. The pathophysiology underlying the epileptic process includes mechanisms involved in initiation of seizures (ictogenesis), as well as those involved in transforming the normal brain into a seizure-prone brain (epileptogenesis) [3]; however, this definition does not explain the biological bases of the epilepsy.

Communication between neurons and their targets depends upon the precise timing of electrical and chemical signals. In the nervous system, short, local signals are necessary to convey appropriate timing information. Defects in such timing events produce problematic errors in the final physiological response. Excitability from individual neurons may arise from structural or functional changes in the postsynaptic membrane, alterations in the type, number, and distribution of voltage- and ligand-gated ion channels or biochemical modification of receptors that increase permeability to Ca^{2+}, favoring development of the prolonged depolarization that precedes seizures.

Voltage-dependent calcium channels are the primary triggers for electrically evoked release of chemical transmitters; therefore, understanding the molecular components and events underlying their regulation is central to the development of a mechanistic picture of key events in neuronal signaling. Because hyperexcitation is the key factor underlying ictogenesis, an excessive excitation may originate from individual neurons, the neuronal environment, or neuronal networks [7]. The excitability of synaptic terminals depends on the amount of excitatory neurotransmitter released (e.g., glutamic acid or activation of glutamergic receptors) as well as of an *insufficient* amount of inhibitor neurotransmitter released (e.g., g-aminobutyric acid, or GABA), and consequently of the final extent of membrane depolarization produced (Figure 3).

Neuronal axons have a resting membrane potential of about −70 mV inside versus outside. Action potentials occur due to net positive inward ion fluxes, resulting in local changes in the membrane potential [13]. Membrane potentials vary with the activation of either ligand- or voltage-gated ion channels, which are affected by changes in either the membrane potential or intracellular

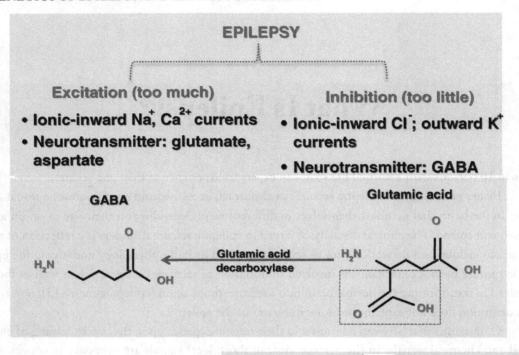

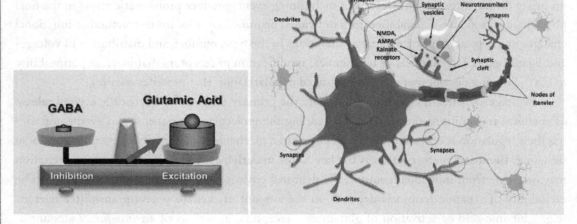

FIGURE 3: (a) The imbalance between the excitatory (high) and the inhibitory (reduced) stimuli are the main mechanism of neuronal membrane depolarization and the onset of seizures. GABA is only synthesized in the nervous system from glutamic acid. (b) Schematic representation of glutamatergic (excitatory) synapses that should be controlled by the inhibitory GABAergic system.

ion concentration. GABA, the principal inhibitory neurotransmitter in the brain, binds postsynaptically to the ionotropic receptor, $GABA_A$, and presynaptically to the metabotropic receptor, $GABA_B$ [14]. Glutamate is the principal excitatory neurotransmitter and binds to both ionotropic and metabotropic types of receptors. The ionotropic receptors as N-methyl-D-aspartate (NMDA), α-amino-3-hydroxy-5-methyl-4-isoxazole propionic acid (AMPA), and kainate (KA) contain subunits whose structure affects the biophysical properties of the receptor. AMPA receptors are the most abundant, followed by NMDA and KA receptors [15]. AMPA receptors have lower glutamate affinity than NMDA receptors, but their faster kinetics account for the fast initial component of the excitatory postsynaptic potential.

Resting Potential

When a neuron is at −70 mV, it is said to be at resting potential. Here, a neuron is not being excited by anything and is basically asleep, awaiting input from other neurons. Resting potential is maintained by the sodium–potassium pump, which constantly pushes Na^+ out of the neuron. The pump operates via a group of sodium–potassium transporters located along the membrane. These transporters push three Na^+ molecules out for every two K^+ molecules it takes in. Resting potential exists when sodium is on the outside of the cell and potassium is on the inside (Figure 4).

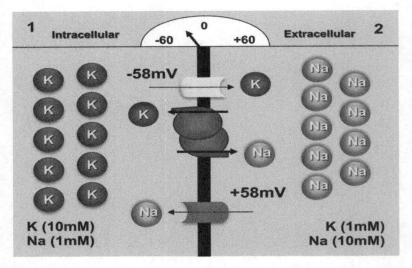

FIGURE 4: Schematic representation of Na^+/K^+ distribution at both sides of the membrane at resting potential and the Golman's equation (JGP 1943).

Action Potential

Action potentials occur when a neuron is excited enough (passes what is called the threshold of excitation) that it depolarizes and the membrane potential shoots from −70 to about 40 mV. After this excitation, the neuron hyperpolarizes, where it will decrease below resting potential and then resume resting potential once the chemicals return to balance. During an action potential, electrostatic pressure and the force of diffusion are key. Ion channels are activated in such a way that Na^+ enters the axon and cell body while, at the same time, K^+ exits the cell body and axon (Figure 5).

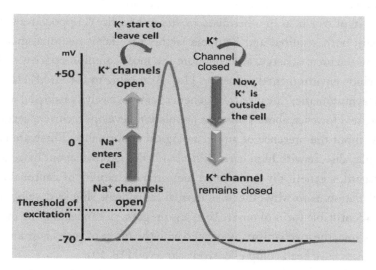

FIGURE 5: Neuronal axons have a resting membrane potential of about −70 mV inside versus outside. Action potentials occur due to net positive inward ion fluxes, resulting in local changes in the membrane potential. Membrane potentials vary with the activation of either ligand- or voltage-gated ion channels, which are affected by changes in either the membrane potential or intracellular ion concentrations.

Are There Any Genetic Bases for Epilepsy?

Epilepsy, for a long time, was not considered a genetic disease, due to sporadic or recurrent seizures that may be caused by many factors such as acquired brain injury (trauma, ischemia, and tumors), metabolic disorders, or congenital malformations. In these epilepsies, called "symptomatic" according to the ILAE classification, a cause (not genetics) can relatively easily be attributed as being responsible for the disease. Moreover, for centuries, epilepsy was considered a "sacred disease," many attributing it to a supernatural role, often demonic. However, in the early Hippocratic writings, epilepsy is considered neither divine nor more sacred than any other disease, one that has a natural cause, and the origin of which, as in other diseases, lies in heredity (Hippocrates 470–410 bc).

Undoubtedly, both acquired and heritable factors may favor neuronal membrane depolarization and initiate seizure discharge, and therefore the most common epilepsy syndromes can be polygenic and include environmental influences [16]. According to the ILAE classification, in the epilepsies called "symptomatic," a cause (not genetics) can be easily considered as the responsible factor for the disease. However, about 1% of the population develops recurrent seizures for no obvious reason and without the presence of any neurological abnormality. These are called idiopathic epilepsies, a complex disease with high heritability, but little is known about its genetic architecture [3, 17]. In this regard, a genetically determined increased excitability of neuronal circuits provides an attractive explanation as to why otherwise normal individuals should develop unprovoked seizures without an identifiable focus of onset. To assess progress in connecting the molecular genetics of epilepsy to the clinic, the mechanisms associated with the range of genes now known to be related to epilepsy syndromes have been comprehensively surveyed [18, 19].

An understanding of the basic mechanisms is fundamental to the design and application of efficacious therapeutics. The inheritance of epilepsy was initially suggested by the observation of familial aggregation, i.e., the fact that relatives of epileptics are often also affected by the disease. Compared to the general population, the risk of epilepsy increases 2 to 4 times in first-degree relatives. However, almost 75% of epileptic people have no affected relatives, and only 15% of IGEs and 12% for cryptogenic focal epilepsies have a familiar epilepsy history. Interestingly, several studies

TABLE 1

EPILEPTIC SYNDROME	MUTATION IN	AFFECTED GENES
(a) With ion-channel dysfunction		
Generalized epilepsy with febrile seizures plus	Sodium channels (β subunit)	*SCN1A/SCN2A/SCN1B*
Benign familial neonatal convulsions	Potassium channels	*KCNQ2, KCNQ3*
Autosomal dominant nocturnal frontal lobe	nAChR	*CHRNA4/CHRNB2*
Juvenile myoclonic	nAChR	*CHRNA7*
IGE	Chloride channels	*CLCN2*
IGE and GEFS+	GABAA receptors	*GABRG2/GABRA1*
Non-ion channel genes in idiopathic epilepsy		
Autosomal dominant lateral temporal lobe epilepsy	Function unknown	*LGI1*
Febrile seizures	G-protein-coupled receptors	*MASS1/VLGR1*
(b) Developmental abnormalities		
Tuberous sclerosis	Hamartin or tuberin	*TSC1, TSC2*
Neurofibromatosis	Neurofibromin	*NF1*
Periventricular nodular heterotopias	Filamin 1	*FLN1*
Miller–Dieker syndrome, isolated lissencephaly	β subunit of PAF acetylhydrolase	*LIS1*
X-linked lissencephaly Doublecortex	Doublecortin	*DCX*

TABLE 1 (*continued*)

EPILEPTIC SYNDROME	MUTATION IN	AFFECTED GENES
(c) *Progressive neurodegeneration/progressive myoclonus epilepsies*		
Huntington's disease	Huntington	*HD*
Fragile X syndrome	FMRP	*FMR1*
Lafora disease	Laforin	*EPM2A/EPM2B (NHLRC1)*
Unverricht–Lundborg disease	Cystatin B	*CSTB*
(d) *Abnormal cerebral metabolism*		
Neuronal ceroid lipofuscinoses and lysosomal proteins	Palmitoyl thioesterase	*CLN1–CLN8*
Myoclonic epilepsy with ragged red fibers	Respiratory chain	*MTTK/MTTL1*
MELAS (respiratory-chain defects)	Mitochondrial DNA mutations	*MTTS1*
	Mitochondrial tRNA (Leu(UUR))	
Leigh disease (respiratory-chain defects)	Mitochondrial DNA mutations	*SURF1*
Inherited metabolic disorders		
Organic acid, amino acid, glycogen storage		
Lysosomal disorders, etc.		

Adapted from "Molecular genetics of human epilepsies" by Bate and Gardiner [148] and "Genetic mechanisms that underlie epilepsy" by Steinlein [149].

have provided evidence for the existence of genetic factors in epilepsy. For example, the concordance rates were higher in monozygotic twins than dizygotic, for both generalized epilepsy (82% versus 26%) and partial (36% versus 5%) [20]. Furthermore, first-degree relatives of patients with idiopathic epilepsy have a roughly two- to three-fold elevated risk of being affected [21]. Because a Mendelian inheritance pattern has been verified in very few epileptic families, other genetic effects may play a role as in several sporadic epilepsies that can be caused by *de novo* mutations (as seen in severe myoclonic epilepsy in children, *SCN1A* gene), or also by somatic mutations in critical regions of the brain [22].

Despite the fact that the genetic etiology in generalized epilepsies is now widely accepted, the focal epilepsies are currently attributed to environmental factors such as damage at birth, infections, head trauma, postnatal, or brain lesions (tumors or vascular damage) [23]. The existence of a family history in patients with partial epilepsy is rare, and it can be explained because, by chance, at least 1 in 20 individuals will have an epileptic-type crisis, perhaps a focal crisis in some point in their life, and they have relatives with epilepsy and status epilepticus [24, 25]. Furthermore, it has come to be believed that there is a basal predisposition to epilepsy that it is heritable, although the syndrome depends on external factors acquired prenatally or postnatally [26]. Meanwhile, families with several affected members may be the consequence of shared exposure to environmental factors or common patterns of behavior such as diet, and not genetic factors. We now know that partial and/or generalized epilepsies could also be genetically determined, and the risk of epilepsy is increased in the relatives of those affected patients. However, as evidenced by twin studies, this risk is greater in the case of generalized epilepsies [27].

As in other common diseases, many of the epileptic syndromes, both generalized and partial, can be explained with a model of multifactorial or complex inheritance in which each gene contributes a small effect, but that gene or variant alone is unable to produce the phenotype. In addition, environmental factors may also play an important role [28]. In these cases, the correlation between genotype and phenotype is relatively weak compared to single gene disorders. This suggests that in the case of a multifactorial model or complex inheritance, and/or inclusion of acquired factors, other not yet studied genes will be linked to the development of epileptic disorders. Although monogenic syndromes represent a small percentage of the total impact of epilepsy, they have become very important for elucidating the molecular mechanisms responsible for epilepsy, and in some cases, single gene mutations can cause both generalized and partial epilepsies, indicating that in epilepsy a loss of relationship of genotype to phenotype can be the rule and not the exception. Thus, the phenotypes associated with some genes can be remarkably variable, even within a family where individuals have the same point mutation.

Although the presence of a mutant allele should be sufficient to cause the manifestation of the epileptic phenotype, several studies show that the penetrance of these mutations usually are not complete, and the age of onset and severity of the phenotype can be variable among families, suggesting that the expression of genes involved in epilepsy can be modulated by additional still unidentified genetic and/or environmental factors [22].

The general epilepsy with febrile seizures plus is a clear example of a phenotype associated with mutations in one gene whose expression can vary even among individuals of the same family who share the same mutation, underscoring the importance of modifying factors. Since the pioneering reports from Berkovics' group, as mentioned above, showing genetic association studies in epilepsy, several pieces of evidence have accumulated indicating that all modes of inheritance (autosomal, X chromosomal, mitochondrial, and complex inheritance) are found in epilepsies. Mutations in genes that code for ion channels or their accessory subunits have been associated with idiopathic epilepsies, and they belong to either the class of voltage-gated ion channels involved in action potential generation and control or the class of ligand-gated ion channels mainly involved in synaptic transmission. In this regard, mutations in voltage-gated potassium, sodium, and chloride channels, as well as in ligand-gated acetylcholine and GABA subunit A (GABA$_A$) receptors have been identified to be related with the cause of different forms of idiopathic epilepsy (Table 1).

Ligand-gated Ion Channels

Mammalian brain ligand-gated ion channels fall into two major superfamilies, the cys-loop receptors (GABA-A, glycine, nicotinic, cholinergic, and 5-HT3 receptors) and the glutamate ionotropic receptors (AMPA, kainate, and NMDA receptors). Additionally, there are purinergic receptors and the ATP-gated P_2X channel. In all cases, binding of agonist to these receptors induces a conformational change that opens the channel. The cys-loop receptors vary in their ion selectivity. GABA-A and glycine receptors are permeable to Cl^- and HCO_3^-, whereas nicotinic cholinergic receptors are permeable mainly to Na^+ and K^+ and also to Ca^{2+}. The ionotropic glutamate receptors are also cation permeable, with significant variation in the extent of Ca^{2+} permeability.

Acetylcholine Receptors *CHRNA4*, *CHRNB2*, and *CHRNA2*

Mutations of the nicotinic acetylcholine receptor (AChR) subunits, that alter amino acids in the transmembrane domains of the AChR proteins, are associated with ADNFLE, a rare epileptic syndrome characterized by partial seizures from the frontal lobe that occur during light sleep; however, only 12% of these patients have mutations on the acetylcholine receptors [29, 30]. The AChR, like glycine, $GABA_A$, and serotonin receptors, is part of the superfamily of homopentameric and heteropentameric ligand-gated ion channels. It is composed of four homologous membrane-spanning subunits forming a cation-selective ion channel, and each subunit contains a large hydrophilic extracellular N-terminus, four putative transmembrane domains (M1–M4), a large cytoplasmic loop between the M3 and M4 and a short extracellular C-terminus (Figure 1).

The M1 and M2 domains form the ion channel pore, whereas the M3 and M4 domains have the largest contact with the membrane lipids and are distant from the ion channel pore and the ligand-binding sites. It was demonstrated that ADNFLE-associated mutations in a4 and b2 subunits reduce the calcium dependence of the acetylcholine response, enhancing excitatory neurotransmitter release during sleep, triggering seizures. Four different mutations in the *CHRNA4* gene have been described in this disorder and three different mutations in the *CHRNB2* gene, which include a functional effect consistent with increased susceptibility to seizures [21]. The nAChR subunits have chromosomal locations as listed in Table 2.

TABLE 2: Chromosomal location of different genes encoding the nAChR subunits

nAChR SUBUNITS	CHROMOSOMAL LOCATION
CHRNA2	8p21
CHRNA3/CHRNA5/CHRNB4	15q24
CHRNA4	20q13.3
CHRNA7	15q14
CHRNB2	1p21.1-q21
CHRNAB3	8p11.22

Adapted from "Molecular genetics of human epilepsies" by Bate and Gardiner [148].

Ion Channels

Ion channels exercise a very tight temporal control over the permeation reaction, as opening a pore across the membrane can be energetically costly. For example, a neuron has a volume of approximately 10^{-12} mL and intracellular K^+ concentration of 100 mM (about 10^8 ions). If ions permeate through a K^+ channel at the rate of 10^7/s, ~10 channel molecules could drain the cell of K^+ in a second. Thus, to be effective in altering the membrane potential—the currency of neuronal signaling—channel opening must be brief and fast. In this context, various ways of turning on (gating) and off (inactivation) in response to external signals or according to in-built mechanisms have been developed, including the self-inactivating property of ion channels, an unprecedented mechanism among biological structures. Additionally, the interaction of channels with cytoplasmic proteins can also trigger conformational changes that modulate the mentioned K^+ channel activity.

Sodium Channels

Voltage-gated sodium channels consist of one large a subunit and the small accessory b subunits (β1–β4), which modulate channel kinetics. The α subunit contains four associated domains (D1–D4) forming the ion pore, and β subunits have a single extracellular IgG loop and a short intracellular C-terminus (Figure 6a–c).

Voltage-gated sodium channels are essential for the initiation and propagation of action potentials in neurons. The space within the membrane forms a sodium-permeable pore, through which sodium ions flow down a concentration gradient during propagation of an action potential.

The first connection between sodium channel alterations and epilepsy was the discovery of a β1 subunit mutation in a large Australian family with generalized epilepsy with febrile seizures plus (GEFS+) (OMIM 604233), a dominantly inherited epileptic syndrome, characterized by febrile seizures in childhood that can progress to generalized epilepsy in adults as well as to an early-onset absence epilepsy [32, 33]. A missense mutation, C121W, in the extracellular domain of the β1 subunit has been described in heterozygous affected family members. Missense mutations in highly evolutionarily conserved amino acid residues T875M or R1648H were described in the α subunit of Na^+ channel in heterozygous affected individuals from two different families, respectively [34]. In addition to these mentioned mutations in the *SCN1A* gene related with GEFS+, a new mutation of *SCN1A* was described in other 7 patients with severe myoclonic epilepsy of infancy (SMEI). SMEI is a rare disorder characterized by generalized tonic, clonic, and tonic–clonic seizures that are initially induced by fever and begin during the first year of life.

Later, patients also manifest other seizure types, including absence, myoclonic, and simple and complex partial seizures. As described in GEFS+, SMEI patients are heterozygous, and interestingly, in the 90% of affected children, mutations emerged *de novo* [35].

Different mechanisms of mutations of the neuronal sodium channel *SCN1A/SCN2A* have been described in patients with epilepsy that include [36]

(a) missense mutations of *SCN1A* (n = 12) related with GEFS+
(b) truncation mutations of *SCN1A* (n = 58) identified in SMEI patients

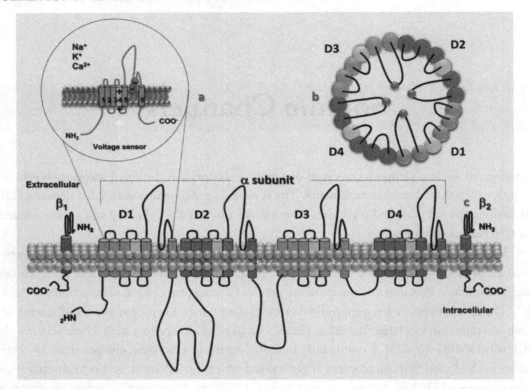

FIGURE 6: Sodium channel. (a) The transmembrane segments are highly conserved through evolution, and the Na⁺, K⁺, and Ca²⁺ channels share the very similar structures. (b) The four homologous domains (D1–D4) of the α subunit are represented in different colors. The transmembrane segments associate in the membrane to form a Na^+-permeable pore lined by the re-entrant S5–S6 pore–loop segments. (c) b1 subunits are represented at the sides of the α subunit. Mutations on the Na^+ channels induce LOF effects.

(c) missense mutations of *SCN1A* in patients with SMEI (n = 47), intractable childhood epilepsy with generalized tonic–clonic seizures (ICEGTC) (n = 7) or infantile spasms (n = 1)

(d) mutations of *SCN2A* in patients with benign familial neonatal–infantile seizures (BFNIS) (n = 5), GEFS+ (n = 1), or SMEI (n = 1)

The large number of *de novo SCN1A* mutations in children with SMEI demonstrates the importance of considering mutation in the etiology of neurological disease, even in the absence of a positive family history. One feature of *SCN1A* mutations is the haploinsufficiency resulting in a quantitative reduction of gene expression to 50% of normal levels, and unlike *SCN1A*, *SCN2A* appears not to exhibit haploinsufficiency [36]. To date, around 700 *SCN1A* mutations are now documented as being associated with seizures, making this the most commonly mutated of the known genes for monogenic epilepsy conditions. Broad phenotypic expression is the hallmark of mutations in this gene ranging from febrile seizures and GEFS+ to Dravet syndrome [35, 36], which generally is accompanied with loss of function (LOF). Most cases of *SCN1A*-associated Dravet syndrome are due to *de novo* mutations but 5% appear in GEFS+ families where the associated phenotypes in other family members are less severe. In GEFS+ mutations (all missense mutations so far) a change in expression of the sodium channel through a variety of electrophysiological mechanisms suggests either LOF in some situations, or gain of function in other situations.

Interestingly, the D1866Y mutation in the *SNC1A* gene, does not alter the voltage dependence of sodium channel activation; however, due to its location at the C-terminal a subunit, an intracellular interaction domain appears to be required, in combination with the extracellular interaction domain to form the stable α/β complex. The b1 subunits contain multiple extracellular and intracellular a interaction domains and all of these domains must be intact for complete β1-mediated modulation of α to occur. Functional studies have identified the cytoplasmic C-terminus of several sodium channel isoforms as playing a modulator role in the voltage dependence of fast inactivation. Therefore, the D1866Y mutation weakens this interaction, demonstrating a novel molecular mechanism leading to seizure susceptibility, due to D1866Y mutation altering sodium channel rapid voltage-dependent "inactivation," which is critical for proper sodium channel function. This mutation results in a positive shift in the voltage dependence of sodium channel inactivation leading to neuronal hyperexcitability. Since mutations in either *SCN1A* or β1 can result in GEFS+, it is not surprising that impaired interaction between the two subunits can also cause the disease [37].

In addition, Dravet syndrome can arise as "pertussis vaccine encephalopathy" secondary to pertussis vaccine administration. The risk of seizures and other acute neurological illnesses after immunization with pertussis-containing vaccines was recently reported [38]. There is no evidence that vaccinations with pertussis toxin before or after disease onset affect outcome of Dravet syndrome; however, these vaccinations might trigger earlier onset of Dravet syndrome in children who, because of an *SCN1A* mutation, are destined to develop the disease [38]. All these findings indicate that environmental factors could precipitate the onset of an epileptic disease genetically determined, as well as their clinical severity.

Meanwhile, and irrespective of sodium channel mutations, a Dravet-like syndrome secondary to mutations in protocadherin 19 (*PCDH19*) can be responsible for a Dravet syndrome that mainly affects the females. One of the earliest stages of brain morphogenesis is the establishment of the neural tube during neurulation. *PCDH19*, a member of the cadherin superfamily, is expressed in the anterior neural plate and is required for brain morphogenesis. Experimental interference of *PCDH19* function leads to a severe disruption in early brain morphogenesis. However, despite these pronounced effects on neurulation, axial patterning of the neural tube appears normal [39]. The mutations in *PCDH19* identified to cause EFMR are mostly found in exon 1 of the gene, which is responsible for the extracellular cadherin domain of that protein. This means that when a mutation occurs that affects the shape of the protein, it is no longer able to bind properly to the other cells using that cadherin domain.

A hemizygous deletion on chromosome Xq22.1, encompassing the *PCDH19* gene, was found in one male patient, and later, *PCDH19* mutations were reported to be the causative gene for female-limited epilepsy and cognitive impairment (EFMR), a disorder characterized by seizure onset in infancy or early childhood and cognitive impairment, which is found only in females in multigenerational families [40].

PCDH19 is a 1148-amino acid transmembrane protein belonging to the PCDH19 δ2 sub-class of the cadherin superfamily, which is highly expressed in neural tissues and at different developmental stages. The δ-PCDH19s mediate cell–cell adhesion *in vitro* and cell sorting *in vivo* and regulate the establishment of neuronal connections during brain development, the precise function of the protein remain so far unknown. However, it has been postulated that random inactivation of one X chromosome in mutated females generates tissue mosaicism (i.e., coexistence of *PCDH19*-positive or *PCDH19*-negative cells), altering cell–cell interactions. Normal individuals and mutated males, who are homogeneous for *PCDH19*-positive or *PCDH19*-negative cells, respectively, would not develop the disease, but this normal cell–cell interaction, could be altered in the presence of the mentioned mosaicism. For this reason, in female cases of Dravet syndrome without mutations in the *SCN1* gene, the study of the *PCDH19* gene is mandatory [41, 42].

Potassium Channel Subunit Mutations with LOF Effects

Potassium channels are a diverse and ubiquitous family of membrane proteins present in both excitable and nonexcitable cells, and over 50 human genes encoding various K^+ channels have been identified. K^+ channels are membrane-spanning proteins that selectively conduct K^+ ions across the cell membrane along its electrochemical gradient at a rate of 10^6 to 10^8 ions/s. Several models of K^+ channels have been described according to the presence of six, four, or two putative transmembrane segments. These include a voltage-gated K^+ channel containing six transmembrane regions (S1–S6) with a single pore (Figure 7), an inward rectifier K^+ channel containing only two transmembrane regions, and a single-pore (not shown), and two-pore K^+ channels containing four transmembranes with two-pore regions (not shown). The voltage-gated K^+ channels are composed of four subunits each containing six transmembrane segments (S1–S6) and a conducting pore (P) between S5 and S6 with a voltage sensor (positive charge of amino acid residues) located at S4, and also can include an auxiliary β subunit (Kvβ), which is a cytoplasmic protein with binding site located at the N terminus of the α subunit (Figure 7).

Defects in the genes encoding the neuronal potassium channel subunits *KCNQ2* and *KCNQ3* are responsible for benign familial neonatal epilepsy (BFNE) [43–45], an autosomal, dominantly inherited epilepsy of the newborn. BFNE is characterized by the transient nature of unprovoked, generalized, or multifocal seizures, sometimes lateralized motor seizures, with onset at 3 days of life and ending by 3 months of age, and the course of the disorder is usually self-limiting. Despite their benign nature, several patients with BFNC have been described in which the seizures did not respond well to AEDs and resulted in delayed development or psychomotor retardation [46]. Perhaps these patients have a second unrecognized condition or the presence of certain risk factors in combination with the BFNC mutation could be responsible for the unfavorable outcome. This short time of expression of the convulsive phenotype can be explained by the expression of compensatory mechanisms such as GABA receptors, which mature as the neonate develops [47]. Because K^+ channels play fundamental roles in the regulation of membrane excitability, it is to be expected that both genetic and acquired diseases involving altered functioning of neurons, smooth muscle, and cardiac cells could arise subsequent to abnormalities in K^+ channel proteins.

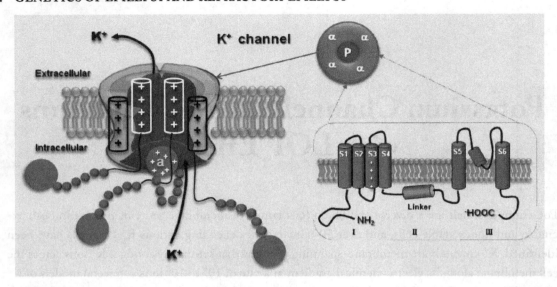

FIGURE 7: K$^+$ channel. It is composed of eight subunits (four α subunits, which span the cell membrane, and four β subunits that lie just inside the membrane). The α subunits have three components, the uppermost being the pore-forming domain. The theory of channel inactivation by the "ball-and-chain system" (shown in red) indicates that for inactivation to occur, a positively charged inactivation particle ("a" ball) has to pass through one of the lateral windows and bind in the hydrophobic binding pocket of the pore's central cavity. This blocks the flow of potassium ions through the pore. There are four balls and chains to each channel, but only one is needed for inactivation.

Genetically linked diseases of the cardiac, neuronal, neuromuscular, renal, and metabolic systems involving members of voltage-gated K$^+$ channels, inward rectifiers, and channel-associated proteins have been decribed [48]. Voltage-gated potassium channel genes *KCNQ2* or *KCNQ3* are located on chromosome 20q13.3 and 8q24, respectively, and encode subunits of the M-channel, a powerful controller of neuronal repetitive firing found ubiquitously in the brain. BFNE cases due to the mutation of *KCNQ3* are much rarer, with only four described so far. Instead, mutations of the *KCNQ2* gene include missense, nonsense, splice site, and frameshift changes as well as altered copy number, and the total number of known pathogenic mutations in *KCNQ2* is approaching 100, making it currently the second most commonly mutated gene for monogenic epilepsies after *SCN1A* [49]. The haploinsufficiency of *KCNQ2* is the main accepted mechanism for relaxation of seizure inhibition, where the size of the molecular defect does not generally amplify the clinical severity, which is uniformly benign in 85% of cases of BFNE. In the remaining 15% of patients with BFNE secondary to *KCNQ2* mutations, progressive severe encephalopathies with intellectual disability have been reported [50–52].

Voltage-gated Calcium Channel Mutations and Genetic Susceptibility

Voltage-gated calcium channels are the mediators of calcium entry into neurons in response to membrane depolarization. Thus, the activation of calcium-dependent enzymes, gene expression, the release of neurotransmitters from presynaptic sites, and the regulation of neuronal excitability are mediated by the influx of Ca^{2+} via these channels. Because the nervous system expresses different calcium channels with unique cellular and subcellular distributions and specific physiological functions, they have been classified as low voltage-activated (LVA) calcium channels (i.e., T-type channels) and high voltage-activated (HVA) channels. Whereas LVA channels are activated by small depolarizations near typical neuronal resting membrane potentials and are key contributors to neuronal excitability, HVA channels require larger membrane depolarizations to open and can be further subdivided into L, N, R, P, and Q types according to specific pharmacological and biophysical properties (Table 3). It is important to note that in the context of neurotransmitter release, N-type channels tend to support inhibitory neurotransmission, whereas the P/Q-type channels have more frequently been linked to the release of excitatory neurotransmitters. The molecular structure of HVA voltage-gated Ca^{2+} channels indicate that they are heteromultimers formed through association of different subunits (α_1, β, α_2–δ, and γ) (Figure 8).

The α_1 subunits comprises four major transmembrane domains that are structurally homologous to those found in voltage-gated sodium and potassium channels, with intracellularly localized NH_2 and COOH termini (see Figure 6a) and the α-helix S_4 segment, containing positively charged arginine and lysine residues every three to four amino acids. This segment can translocate within the membrane in response to changing membrane potential and so acts as a voltage sensor mediating the channel opening. Between the S5 and S6 segments, four glutamic acid residues form a ring of negative charge, that allows more selectivity for divalent cations. The α_2–δ subunit is cleaved into a membrane-spanning δ-peptide and extracellular α_2-peptide, and it is known that its association with gabapentin mediates analgesia [53].

The rhythmic activity that is observed in EEG recordings during an epileptic seizure is a reflection of cortical and thalamic network interactions. Consequently, T-type channels have always been likely candidates due to their eminent presence in cortical and thalamic structures and their

TABLE 3: Neuronal α subunits		
HVA	L type	Ca$_v$1.2, Ca$_v$1.3, Ca$_v$1.4
	P/Q type	Ca$_v$2.1
	N type	Ca$_v$2.2
	R type	Ca$_v$2.3
LVA	T type	Ca$_v$3.1 (thalamocortical neurons)
		Ca$_v$3.2, Ca$_v$3.3 (thalamic reticular nucleus)

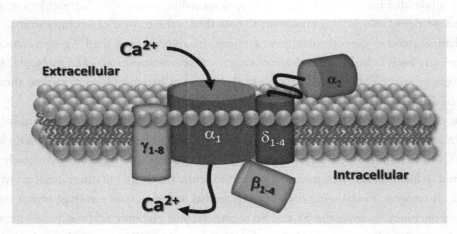

FIGURE 8: Subunit assembly and subtypes of voltage-gated calcium channels. The high voltage-activated calcium channel complex consist of a main pore forming α1-subunit (blue) plus ancillary subunits (β, γ, and α2-δ). The γ subunits comprises four transmembrane domains with intracellular NH$_2$ and COOH termini, but the mutual sites of interaction with the α1 subunit remain unknown. Low voltage-activated calcium channels may be formed by a α1 subunit alone.

established physiological role in modulating neuronal firing. A number of missense mutations have now been identified in the Cav3.2 calcium channel gene in patients diagnosed with childhood absence epilepsy and other forms of IGE.

The missense mutations on Cav3.2 channel are consistent with a gain of function, as observed in some of the mutations resulting in a hyperpolarizing shift in the voltage dependence of activation, whereas others resulted in increased channel availability due to decreased steady-state inactivation. Several mutations in voltage-gated calcium channels have been reported related with genetic susceptibility to epileptic syndromes such as childhood absence epilepsy or IGE. Mutations of the *CACNA1H* gene on chromosome 16p13.3 have been found in several patients with childhood absence epilepsy, suggesting that *CACNA1H* might be a susceptibility gene that is involved in the pathogenesis of IGEs. Similarly mutations of the α_1A-calcium channel subunit gene (CACNA1A) on chromosome 19, were associated with IGEs that include age-related subtypes such as juvenile myoclonic epilepsy, childhood absence epilepsy, juvenile absence epilepsy, and grand mal epilepsy on awakening [54, 55] (Figure 9).

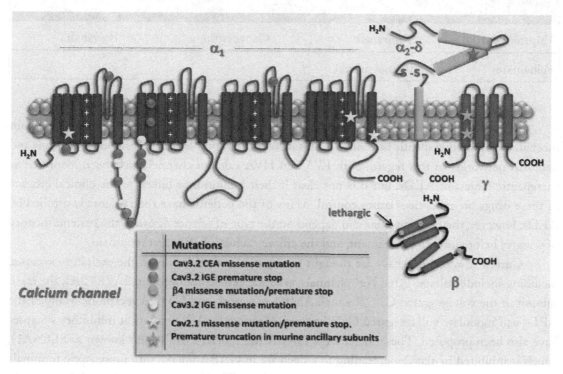

FIGURE 9: Locations of identified mutations linked to epilepsy in the calcium channel α_1 and associated ancillary subunits.

TABLE 4: AEDs categorized by mechanism of action			
SODIUM CHANNEL BLOCKERS		**CALCIUM CHANNEL BLOCKERS**	
Phenytoin	Carbamazepine	Topiramate	Lamotrigine
Oxcarbazepine	Lamotrigine	Ethosuximide	GABA enhancers
Zonisamide	Lacosamide	Benzodiazepines	Tiagabin
	Rufinamide	Vigabatrin	Phenobarbital
GLUTAMATE RECEPTOR ANTAGONISTS		**POTASSIUM CHANNEL OPENERS**	
Topiramate	Felbamate	Retigabine	Diuretics
		Bumetanide	Acetazolamide
MULTIPLE MECHANISMS		**GABAPENTENOIDS**	
Valproic acid	Topiramat	Gabapentin	Pregabalin
Felbamate	Phenobarbital		

However, all these observations remain to be confirmed by more extensive studies. The main mechanism of the therapeutic benefit of AEDs involves elevating seizure threshold for neurons and neuronal networks. In this regard, both LVA and HVA calcium channels have been identified as therapeutic targets of AEDs, but it is not clear if their inhibition is linked to the clinical efficacy of these drugs on epileptic seizures control. Many of the patients have their seizure controlled by AEDs; however, their effectiveness can depend on the type of seizure disorder, the patients' history of seizures before initiating treatment, and the efficacy achieved after first treatment.

Currently available AEDs are thought to target several molecules at the excitatory synapse, including include voltage-gated Na^+ channels, synaptic vesicle glycoprotein 2A (SV2A), the α_2–δ subunit of the voltage-gated Ca^{2+} channel, AMPA receptors, and NMDA receptors. Many of the AEDs can modulate voltage-gated Na^+ channels. Meanwhile, AED targets at inhibitory synapses have also been proposed. These include the GABA transporter GAT1 (also known as SLC6A1), which is inhibited by tiagabine, leading to a decrease in GABA uptake into presynaptic terminals and surrounding glia, and GABA transaminase (GABA-T), which is irreversibly inhibited by vigabatrin (Table 4).

GABA Receptor Subunit Mutations (Chloride Channel)

Because the $GABA_A$ receptor is the major inhibitor of neuronal transmission in the central nervous system (CNS), mutations interrupting the normal export of chloride ions through the $GABA_A$ receptor affect the capability of the neuron to inhibit hyperexcitability. Two of the multiple subunits (*GABRA1* and *GABRG2*) have been implicated in familial epilepsy [55–57]. $GABA_A$ receptors are anion-selective ligand-gated channels that mediate fast synaptic inhibition (Figure 10).

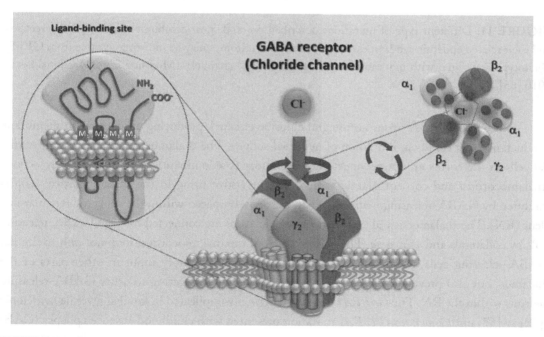

FIGURE 10: Schematic representation of $GABA_A$ receptor and membrane topology (bottom). Each subunit contains four transmembrane segments of typical α1 subunit (left). The M2 transmembrane domains from each subunit (blue) are spatially located to build a selective chloride pore (right).

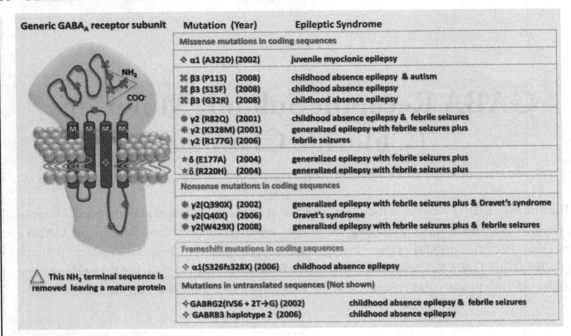

FIGURE 11: Different type of mutations described on α, β, γ, or δ subunit of the GABA$_A$ receptor and their related epileptic syndromes. Several of these syndromes may be indistinguishable from GEFS+ phenotype associated with mutations in voltage-gated Na$^+$ channels. (Modified from Macdonald et al. 2010 [65].)

Binding of GABA opens an integral chloride channel, producing an increase in membrane conductance that results in inhibition of neuronal activity. The thalamus and the layers of pyramidal cells in the cortex are interconnected by excitatory (predominantly glutamatergic) projections (thalamocortical and corticothalamic axons). The repetitive firing in the thalamocortical loop is inhibited by GABA-releasing neurons, which are densely spaced within the thalamic reticular nucleus (RN). The thalamocortical and corticothalamic axons are connected to these GABA-releasing cells by collaterals and *vice versa*. The collaterals from the thalomocortical loop not only excite the GABA-releasing cells, which subsequently increase their inhibitory input in other parts of the thalamus, but also prevent excessive inhibitory stimulus by connecting to other GABA-releasing neurons within the RN. The *GABRA1* gene was originally implicated in familial juvenile myoclonic epilepsy [57] until a *de novo GABRA1* mutation presented with childhood absence epilepsy (CAE) [58], demonstrating a role for shared genetic determinants among IGE subtypes (Figure 11).

The initial GABA$_A$ receptor mutations associated with IGE were found in the γ2 and α1 subunits, consistent with a genetic defect in phasic-synaptic GABAergic inhibition, and at the

GABRD gene encoding the δ subunit, a mutation related with susceptibility locus for IGEs was also reported [59]. A complete mutation map on different subunits and related with several epileptic syndromes has been reported (Figure 11) [60]. The d subunit is present in extrasynaptic and perisynaptic receptors, which mediate tonic inhibition, suggesting that this mechanism plays a role in epilepsy. Once all subunits of GABA$_A$ receptor are assembled, a chloride ion channel is formed. Several positive and negative allosteric regulators—including barbiturates, benzodiazepines (BZDs), and neurosteroids, as well as bicuculline, picrotoxin, and zinc—can modulate the GABA$_A$ receptor currents. Inhibitory postsynaptic currents (IPSCs) are triggered by release of presynaptic GABA, which binds to postsynaptic GABA$_A$ receptors. Interestingly, the δ subunit plays a role for abd receptors in tonic inhibition. The αβδ receptors desensitize more slowly and less extensively than αβγ receptors and have a lower GABA EC50. *GABRA1* and *GABRG2* genes encode the α$_1$

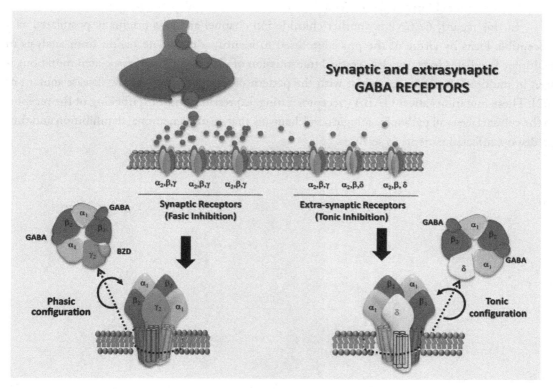

FIGURE 12: The junctions between α and β subunits configure a typical synaptic (phasic) GABA$_A$ receptor (left), whereas the δ subunit replaces the γ subunit in their interaction with the α subunit in a typical extrasynaptic (tonic) GABA$_A$ receptor.

and γ_2 subunits of the GABA$_A$ receptor (respectively), and these subunits co-assemble together with β subunits to make up an abundant receptor subtype expressed in the forebrain. More recently, Lachance-Touchette et al. [61] screened three French Canadian families with IGE for mutations in *GABRA1* and *GABRG2* and reports two previously unreported mutations in *GABRA1*. The functional analysis was consistent with a reduction in cell surface expression of the mature protein in both cases, where the receptors are insensitive to Zn^{2+} and are potentiated by diazepam, important controls to show that the γ_2 subunit is incorporated into GABA$_A$ receptors at the cell surface (Figure 12).

Although a common mutation-mediated loss of protein function was described, the seizure phenotypes of patients are different and include febrile seizures, generalized tonic–clonic (GTC) seizures, and photosensitive seizures. This broad seizure spectrum reinforces the concept of a loss of genotype–phenotype relationship in monogenic epilepsies and suggests that other genetic factors could be critical in defining seizure outcome. Because the GABA receptor is a chloride channel, mutations on other chloride channels could be associated with epileptic syndromes.

In this regard, *CLCN2* is another chloride ion channel and was originally postulated as a susceptible locus by virtue of the procedure used to identify a candidate region from analysis in multiplex families. However, clear vertical transmission of phenotypes with associated mutations is seen in multiplex families, consistent with the pattern observed for monogenic disease mutations [62]. These mutations affect GABA$_A$ receptor gating, expression, and/or trafficking of the receptor to the cell surface—all pathophysiological mechanisms that result in neuronal deinhibition and thus predispose affected patients to seizures.

Copy Number Variants and Comorbidities

Until recently, our concept of the human genome was of 23 pairs of chromosomes, each with a continuous, folded strand of the elegant double helix that remains unchanged in gene number from generation to generation. However, deletion or duplication of various stretches of DNA, usually incorporating a number of genes, occurs frequently throughout the genome in healthy subjects. The multifactorial nature of epilepsy includes the never disclosed "predisposing" and "exciting" factors. In this regard, in the epilepsies for which no structural or metabolic cause can be demonstrated, mostly sporadic forms termed idiopathic or cryptogenic, should be analyzed under an *epigenetic* reinterpretation, focusing on the early concepts that seizures are often influenced by inherited and environmental factors.

Perhaps, neurodevelopmental disorders are likely to be the result of a complex interplay between genetic and epigenetic influences, where epilepsy is one of all protagonists. Furthermore, of relevance to epilepsy (and other neurodevelopmental conditions) is that analysis of microarray gene expression data generated by specific human brain regions reveals that gene network structure can vary across anatomical brain regions as well as with age [63]. The recent discovery of the importance of CNVs across a range of neurodevelopmental disorders appears to be similar to a wide range of epilepsies associated with psychiatric comorbidities.

Comorbidities in epilepsy are often attributed to drug treatment or the direct or indirect effects of epileptic disease. More recently, one other explanation of these epidemiological data is that there is an etiology common to epilepsy and its cognitive and behavioral comorbidities.

The bidirectional nature of the association of epilepsy with several neuropsychiatric comorbidities, suggests that may share the same genetic mechanisms. The so-called neurodevelopmental disorder typically refers to the range of common neurological and behavioral conditions that includes schizophrenia, autism, and mental retardation, intellectual disability, could present copy number variants (CNVs) in their pathogenesis. Furthermore, several important studies have demonstrated an association between this class of CNV and epilepsy, autism, schizophrenia, MR/ID, and attention deficit hyperactivity disorder [64].

The neuropsychological (cognitive and behavioral) comorbidities of epilepsy have been recognized for centuries, but the extent to which neuropsychological comorbidities are important in the prognosis and care of patients with epilepsy has received more emphasis in the past two decades. In recent years, several studies have shown clear links between epilepsy and various neuropsychiatric disorders including psychosis and depression, and genetic studies of CNVs have shown that in some cases, the same CNV underpins neuropsychiatric illness and epilepsy. CNVs are invisible on routine karyotyping but are readily revealed using higher definition molecular techniques such as single-nucleotide polymorphism (SNP) microarrays or array comparative genomic hybridization. These techniques allow at least 10 times the resolution of chromosomal structure, and as techniques are further refined to allow still greater resolution, the number of CNVs will undoubtedly increase.

These CNVs result in the usual two copies of a gene changing to a single copy in a heterozygous deletion, microdeletion or more than two copies in duplications, and often have no phenotypic expression, but they can act as rare variants predisposing to complex disease. Larger CNVs are more likely to be associated with disease although size does not directly correlate with gene content [65]. First identified in 0.3% of patients with mental retardation, dysmorphic features, and seizures, the same 15q13.3 microdeletion also observed in 0.2% of patients with schizophrenia [66] was now detected in patients in IGE.

It is not clear how do these microdeletions confer their pathogenic effect. However, haplo-insufficiency of specific genes within a deletion is the favored hypothesis. For example, the most promising suspect in the 15q13.3 microdeletion is *CHRNA7*, which encodes the α7 subunit of the nicotinic receptor.

Susceptibility Genes for Complex Epilepsy

Common idiopathic epilepsies are a heterogeneous group of complex seizure disorders, where the genetic component is mostly polygenic, and each susceptibility allele alone is insufficient to cause seizures but requires the additive or epistatic interaction of other susceptibility alleles. Since 1995, the continuous identification of susceptibility genes affecting transport of ions into the neuron, suggest that progress towards discovering the underlying genetic variation for susceptibility to complex idiopathic epilepsy has only just begun (Table 5) [67].

Interestingly, more recently, it was demonstrated experimentally that seizure susceptibility in genetically epilepsy-prone rat (GEPR) is associated with altered protein expression of voltage-gated calcium channel subunits in inferior colliculus neurons. Furthermore, in these experiments, a single seizure selectively enhanced protein expression of Ca^{2+} channel $\alpha 1A$ subunits in IC neurons of GEPR-3s. Thus, the upregulation of Ca^{2+} channel $\alpha 1D$ and $\alpha 1E$ subunits may represent the molecular mechanisms for the enhanced current density of L and R types of HVA Ca^{2+} channels in IC neurons of the GEPR and may contribute to the genetic basis of their enhanced seizure susceptibility [68].

Neuronal activity is regulated by the concentration of ions in the extracellular and intracellular spaces and the selective flux of these ions across the neuronal membrane. Voltage- or ligand-gated ion channel genes are therefore attractive candidate genes for the epilepsies. It is easy to imagine how mutations in such genes could lead to channel dysfunction, which could alter ion concentrations across the cell membrane, resulting in reduced or increased neuronal excitability. However, it is not so easy to image how an epileptic syndrome could be developed secondary to a very different pathological process. Several questions have been clearly expressed focusing to the future of the genetic predisposition in epilepsy [67]; however, it should also be addressed if an additional potential association between the environmental factors and the described susceptibility alleles could to explain why some patients, but not all of them, develop secondary epilepsies after the same brain insult, including the repetitive uncontrolled seizures (Figure 13). Thus, the nongenetic causes of epilepsy are really nonrelated with genetic causes or genetic epilepsy predisposition.

TABLE 5	
EPILEPSY-RELATED GENE	EPILEPSY SYNDROME
Monogenic channelopathies	
CHRNA4	ADNFLE
KCNQ2	BFNS
KCNQ3	BFNS
SCN1B	GEFS+
SCN1A	GEFS+/(SMEI)
CHRNB2	ADNFLE
GABRG2	CAE/FS/GEFS+
SCN2A	GEFS+?/BFNIS
GABRA1	ADJME
CLCN2	IGE
Monogenic, other than channelopathies	
LGI1	ADPEAF
EFHC1	JME
Complex epilepsies: channelopathies	
CACNA1H	CAE, IGE
GABRD	IGE, GEFS+

Interestingly, approximately 30% of children with epilepsy have autism and/or intellectual or developmental disabilities. The increased association of intellectual and developmental disabilities (IDDs), autistic spectrum disorders (ASDs), and epilepsy should be explained from potential same pathophysiological mechanisms, and it has been proposed that IDDs, ASDs, and epilepsy can be understood as disorders of synaptic plasticity resulting in a developmental imbalance of excitation

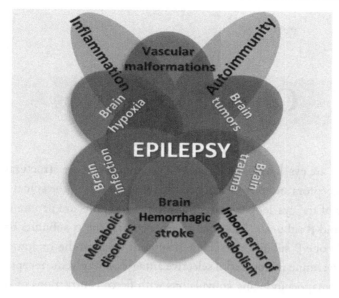

FIGURE 13: Interplay between of more common conditions that develop acquired epilepsy. Despite the observation of an overlapping of several of these mechanisms, only a fraction, and not all, of the affected patients develop an epileptic syndrome from the same primary disease. Is this difference in susceptibility based on any genetic differences between them?

and inhibition. A number of well-known genetic disorders share epilepsy, intellectual disability, and autism as prominent phenotypic features, including tuberous sclerosis complex, Rett syndrome, and fragile X syndrome [69]. In this last syndrome, epilepsy is the most common neurological abnormality, occurring in approximately 20% of children with the disease, and presenting as seizure and EEG abnormalities. In a mice model of fragile X syndrome, increased seizure susceptibility is attenuated by reducing expression of metabotropic glutamate receptor 5 (mGluR5) as well as by use of a receptor antagonist [70].

Glycine Receptors

The glycine receptors are cys-loop ligand-gated Cl⁻ channels that are structurally and functionally similar to GABA$_A$ receptors and play an important inhibitory role in the spinal cord and brainstem. Like other members of the cys-loop superfamily, glycine receptors are pentameric in structure and are composed of α and β subunits. Multiple genes encoding the α subunits have been identified, whereas there is only one β subunit gene. The β subunit anchors the receptor to the cytoplasmic protein gephyrin. Strychnine is a powerful selective antagonist of glycine receptors that binds selectively to glycinergic synapses inducing convulsions with fierce contractions of skeletal musculature without loss of consciousness. No naturally occurring epilepsy syndrome in humans has the "spinal" characteristic observed during the strychnine-induced convulsions, and no human epilepsy syndromes have been associated with glycine receptor mutations. However, at least 29 mutations in the *GLRA1* gene have been found to cause hereditary hyperekplexia (Startle disease). Most of these mutations change single amino acids in the α1 subunit of the glycine receptor protein (Figure 14).

The most common mutation replaces the amino acid arginine with the amino acid leucine at protein position 271 (written as Arg271Leu or R271L). Glycine receptors lack a functional role in the forebrain, so they are unlikely to be a useful AED target. Interestingly, neurotransmitter glutamate that serves to activate all ionotropic glutamate receptors at synapses requires the presence of a coagonist, either glycine or D-serine to activate the NMDA receptors. It is important to notice that several enzymatic deficiency related with glycine metabolism were described with severe seizures or epileptic syndromes.

- Glycine cleavage system
- 3-Phosphoglyceratedehydrogenase
- Phosphoserine aminotransferase
- 3-Phosphoserine phosphatase

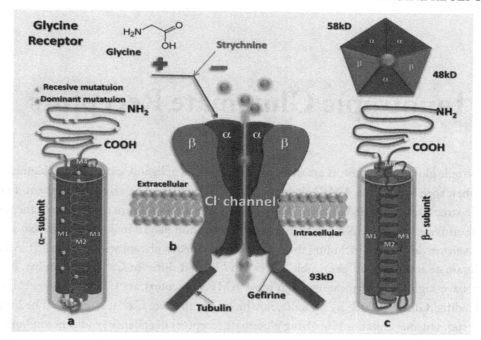

FIGURE 14: Schematic diagram suggests four-membrane spanning domain (M1–M4) topology of GlyR α1 (a) and GlyR β (c) subunits. Point of mutations on glycine receptor are indicated by recessive (yellow circles) and dominant (sky-blue circles) human hyperekplexia.

In glycine cleavage system deficiency as the nonketotic hyperglycinemia, patients with neonatal form develop deep mental retardation, a severe myoclonic and generalized seizure disorder, pronounced axial hypotonia, and spastic quadriplegia. Some patients have hypoplasia of the corpus callosum, cortical malformations, or hydrocephalus with posterior fossa cystic malformation. Patients with the late-onset variant tend to have present seizures, moderate mental retardation, ataxia, hyperactivity, and/or chorea. A few common mutations have been identified.

Ionotropic Glutamate Receptors

The ionotropic glutamate receptors are three receptor families of ligand-gated cation channels identified by their specific agonists AMPA, kainate, and NMDA that selectively activate them. They are tetrameric structures composed of more than one type of subunit that mediate most of the fast excitatory transmission in the CNS and their excessive activation plays a main role in the exitotoxicity as well as seizure discharges, including the epileptic spontaneous phenomena (Figure 15). All ionotropic glutamate receptors are permeable to Na^+ and K^+ but differ in Ca^{2+} permeability. NMDA receptors have high Ca^{2+} permeability, and most AMPA receptors are Ca^{2+}-impermeable, unless they lack edited GluR2 (GluR-B) subunits, in which case, they are Ca^{2+}-permeable. The little evidence for spontaneous mutations involving glutamate receptors in epilepsy syndromes in human are related with juvenile absence epilepsy are associated with a nine-repeat allele of a tetranucleotide repeat polymorphism in a noncoding region of the GluR5 receptor gene (*GRIK1*); however, other studies failed to confirm that mutations in this syndrome.

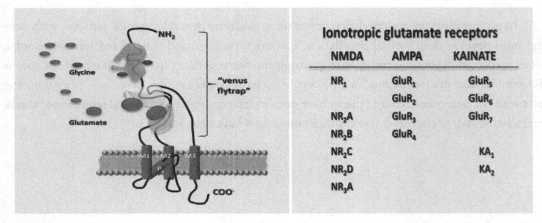

Ionotropic glutamate receptors

NMDA	AMPA	KAINATE
NR_1	$GluR_1$	$GluR_5$
	$GluR_2$	$GluR_6$
NR_2A	$GluR_3$	$GluR_7$
NR_2B	$GluR_4$	
NR_2C		KA_1
NR_2D		KA_2
NR_3A		

FIGURE 15: Extracellular amino terminal domain with the ligand-binding structure named "Venus flytrap." The binding of two glycine and two glutamate molecules is required for channel opening. The functional channel comprises two dimers. (The binding sites of glycine and glutamate are hypothetical.)

G-protein-coupled Receptors

G-protein-coupled receptors have been identified from the sequencing of the human genome but are of unknown function. These receptors have an extracellular N-terminal, an intracellular C-terminal, and seven α-helical transmembrane segments and are divided into three main classes (A–C). Class C includes the mGluRs and the GABAB receptors, which play important roles in controlling excitability at glutamatergic and GABAergic synapses. Both types of receptors share several features, although the mGluRs seem to be mostly homomeric, whereas functional GABAB receptors are heterodimers [71].

Metabotropic Glutamate Receptors

The mGluRs are a family of eight G-protein-linked receptors that fall into three groups defined by their sequence homology, second messenger effects, and common pharmacology. Their distribution within the CNS provides a platform for both presynaptic control of glutamate release and postsynaptic control of neuronal responses to glutamate. In neurons, mGluRs located postsynaptically have excitatory effects mediated by several ion channels, while presynaptic mGluRs have been shown to control synaptic release (Figure 16).

The mGluRs regulate intracellular second messenger systems that take part in a variety of physiological and pathophysiological processes, including development, learning and memory, pain, stroke, epileptic seizures, and schizophrenia, as well as in chronic neurodegenerative diseases such as Alzheimer, Huntington, and Parkinson diseases.

The discovery of the mGluRs over the past few decades has led to a better understanding of the potential long-term consequences of excessive glutamatergic transmission. These G-protein-coupled receptors activate intracellular cascades of events, resulting in long-lasting modifications of cellular and network excitability.

The eight known mGluRs (mGluR1–8) are subdivided into three groups, based on sequence homology, agonist specificity, and associated second messenger systems.

Acquired forms of epilepsy show marked changes in the expression and function of mGluRs; however, there is little evidence to indicate that genetic alterations of mGluRs play a role in the pathogenesis of epilepsy. Experimentally, mice with GluR7 deleted by gene targeting show an increased susceptibility to seizures and increased cortical excitability, and in the kindling model of epilepsy, there is upregulation of postsynaptic mGluRs receptors [71, 73], as well as downregulation of presynaptic mGluRs of hippocampal slices from kindled rats [74] and from patients with complex partial seizures [75]. According with these observations, the upregulation of mGluR5 was also reported after epileptic seizures in the surviving neurons from temporal lobe of patients with refractory TLE, suggesting that this upregulation could also contribute to the hyperexcitability of the hippocampus in these pharmacoresistant TLE patients. This prominent role of mGluR5 in human TLE indicates mGluR5 signaling as potential target for new AEDs [76].

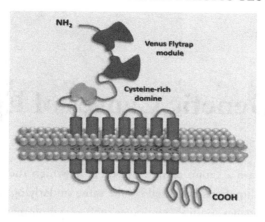

FIGURE 16: Typical membrane topology of class "C" G-protein-coupled receptors. Intracellular C-terminal, extracellular N-terminal, seven transmembrane helices, the cysteine-rich domain, and the "Venus flytrap" module are a common feature of glutamate and GABA metabotropic receptors.

Contrarily, after pilocarpine-induced status epilepticus, a quantitative measurements of group I mGluR proteins revealed selective downregulation of mGluR5 and its membrane anchoring protein, whereas mGluR1 expression remained without changes. Although mGluR1 and mGluR5 are members of the same family, their respective roles in seizure production and epileptogenesis differ. *In vitro* studies have shown that transient activation of group I mGluRs will result in long-lasting, spontaneous ictal discharges and that, while mGluR1 and mGluR5 work together in both the induction and maintenance phases, it is primarily mGluR5 that drives the epileptogenesis in the induction phase and mGluR1 that maintains the ongoing expression of the seizure discharges. Furthermore, the selective downregulation of mGluR5 in the chronic phase in part may explain the increasingly important role recognized for mGluR1 in the ongoing expression of seizure discharges following epileptogenesis [77, 78].

All these data suggest changes in functional expression of these receptors could tend to enhance excitability. Like mGluRs, GABAB receptors are G-protein-coupled receptors, and in the genetic absence epilepsy rats from Strasbourg model of absence epilepsy, there is an upregulation of both GABAB1 and GABAB2 protein in corticothalamic circuits, which may contribute to the seizure phenotype [79]. It is clear that upregulation or downregulation of different genes are not any mutation; however, they could be addressed as genetic and/or epigenetic regulations, with direct consequences on seizures susceptibility, epilepsy development, epileptogenesis, or include pharmacoresistant phenotype as well.

Other Genetic Causes of Epilepsy

Epileptic encephalopathies are a group of rare disorders in which the impairment of cognitive, behavioral, and other brain functions is caused by the same underlying disease process. This heterogeneous group of disorders has multiple etiologies such as symptomatic brain lesions, metabolic causes, and diverse genetic syndromes. In several inherited diseases in which the epileptic seizures are an important, but not the only, clinically prominent feature, the epilepsy is often accompanied by other neurological symptoms, such as mental retardation, dementia, or ataxia. In contrast to the idiopathic epilepsies, the course of these disorders is often not benign and developmental abnormality or irreversible and progressive neuronal cell losses in the brain are produced. These abnormalities include severe alteration on respiratory chain activity, glycogen metabolism, or brain development.

LAFORA DISEASE

It is an autosomal recessive neurodegenerative disorder and is the most common form of adolescent-onset progressive epilepsy, leading to cognitive decline, dementia, and finally death within 10 years of onset. Intracellular periodic acid Schiff-positive inclusion bodies, known as Lafora bodies, in neurons, heart, liver, and muscle are the hallmark of the disease described in 1911 by Lafora. GTC seizures, absences, drop attacks, or focal occipital seizures are frequently observed during adolescence. Homozygous mutations of the *EPM2A* gene on chromosome 6q24 are found in about 70%–80% of the patients. Laforin, the product of the *EPM2A* gene, is an intracellular protein tyrosine phosphatase that acts to oppose the action of tyrosine kinases in cell signaling pathways regulating the levels of phosphotyrosine in cells. Laforin interacts with itself and with the glycogen-targeting regulatory subunit R5 of protein phosphatase 1. A second gene associated with Lafora disease (*EPM2B/NHLRC1*) identified on chromosome 6p22.3, codifies the "Malin," a protein characterized by a zinc finger of the RING type and six NHL-repeat protein–protein interaction domains.

NEURONAL CEROID LIPOFUSCINOSES

The neuronal ceroid lipofuscinoses (NCL) are a group of neurodegenerative encephalopathies characterized by psychomotor deterioration, visual failure, seizures, and the accumulation of autofluorescent lipopigment in neurons and other cell types. Five types of progressive myoclonus epilepsy classified as classical late infantile or CLN2 (Jansky–Bielschowsky disease), juvenile or CLN3 (Spielmeyer–Vogt–Sjögren or Batten disease), adult or CLN4 (Kufs or Parry disease), late infantile Finnish variant or CLN5, and late infantile variant or CLN.

UNVERRICHT–LUNDBORG DISEASE

Also named as progressive myoclonus epilepsy type 1 (EPM1) or Baltic or Mediterranean myoclonus epilepsy, Unverricht–Lundborg disease is an autosomal recessive neurodegenerative disorder that is characterized by progressive, stimuli-sensitive myoclonic jerks, and GTC seizures. In most patients, a dodecamer repeat that is located upstream of the initiation codon of the CSTB (cystatin B, also stefin B) gene was detected. CSTB is intracellular inhibitor of cysteine proteases and is likely to protect the organism from proteolysis by controlling the activities of endogenous proteases. CSTB-deficient mice develop a typical feature of Unverricht–Lundborg disease; however, the convulsive mechanisms observed in this disease are not known.

X-LINKED LISSENCEPHALY/DOUBLE CORTEX SYNDROME

The most common type of lissencephaly is classical lissencephaly, or LIS1, which is characterized by a smooth cerebral surface, a thick cortex, and no other brain malformations. X-linked lissencephaly/double cortex syndrome (XLIS) is an X-chromosomal, dominantly inherited neuronal migration disorder. The affected individuals have epilepsy and mental retardation of variable severity. XLIS is caused by a mutation in the doublecortin (*DCX*) gene, which is located in chromosome Xq22.3-q23 and encodes a microtubule-associated protein that is expressed in migrating neuroblasts. The abnormal microarchitecture of the cortex with aberrant neuronal network can generate focal or secondarily generalized cortical hyperexcitability (Table 6).

TABLE 6		
GENE	CHROMOSOME	PHENOTYPE
RELN	7q22	Lissencephaly and cerebellar hypoplasia
LIS1 (*PAFAH1B1*)	17p13.3	Lissencephaly type 1/Miller–Dieker syndrome
YWHAE (14-3-3ε)	17p13.3	Miller–Dieker syndrome
DCX (*XLIS*)	Xq22.3-q23	Lissencephaly type 1/double cortex syndrome
ARX	Xp22.13	Lissencephaly and genital abnormalities

Mitochondrial Inheritance and Myoclonic Epilepsy with Ragged Red Fibers

The mitochondria has a circular DNA molecule, 16,569 bp long with up to 10 copies per mitochondrion, and therefore the mitochondrial genome (mtDNA) can reach up to several hundred copies per cell. The named "mitochondrial encephalomyopathies" are a genetically and phenotypically heterogeneous group of disorders. Mutations in the maternally inherited mitochondrial genome or by mutations in the nuclear DNA codifying different proteins related with mitochondrial function. mtDNA encodes two ribosomal RNAs, 22 transfer RNAs, and 13 messenger RNAs encoding components of the inner mitochondrial membrane respiratory chain. To date, more than 200 disease-causing mutations of mtDNA are known, and include myopathies, encephalopathies, cardiomyopathies, and various multisystem disorders. The variability of the clinical phenotypes in mitochondrial disorders is mostly due to the heteroplasmy of mtDNA, that is, nonuniform distribution of mitochondria in different tissues and the coexistence of mutated and wild-type mtDNA within the same cell organelle.

Two diseases have been described where CNS involvement is manifested in part as epilepsy caused by point mutations in mitochondrial transfer RNA genes. In both diseases, maternal transmission was documented.

MYOCLONIC EPILEPSY WITH RAGGED RED FIBERS

One of the best-known mitochondrial syndromes is myoclonic epilepsy with ragged red fibers (MERRF), which is characterized by myoclonus, epilepsy, ataxia, muscle weakness, hearing loss, and elevated serum lactate and pyruvate levels. Muscle biopsy typically shows "ragged red" fibers and paracrystalline inclusions in subsarcolemmal mitochondria. The patients are heteroplasmic. Both normal and mutated mtDNA populations are found. Variability of the clinical phenotype appears to depend on the amount and tissue distribution of mutant mtDNA in each individual.

The MERRF syndrome can be caused by mutations in at least two different mitochondrial genes, *MTTK* and, less often, *MTTL1*. *MTTK* and *MTTL1* code for the transfer RNAs for lysine and leucine, respectively. Mutations in these genes result in a severe defect in the translation of mitochondrial genes, causing multiple deficiencies in the cell's respiratory chain. In MERRF, the impairment of mitochondrial protein synthesis mainly affects the cytochrome c oxidase complex (COX). The broad spectrum of possible clinical features in MERRF is due to the wide range of cellular pathways and functions in which mitochondria are involved. The oxidative phosphorylation defect results in cell dysfunction and apoptosis in tissues that are sensitive to decreased ATP production. Mitochondrial defects therefore manifest primarily in organs with a high energy demand, such as brain and muscle. A possible cause of seizures in mitochondrial encephalomyopathies could be an imbalance of excitatory and inhibitory mechanisms within neuronal networks that is caused by functionally impaired neurons.

MITOCHONDRIAL ENCEPHALOMYOPATHY WITH LACTICACIDOSIS AND STROKE-LIKE EPISODES

Mitochondrial encephalomyopathy with lacticacidosis and stroke-like episodes (MELAS) is another mitochondrial encephalopathy with epilepsy, where an A-to-G transition at nucleotide 3243 was reported. Again, heteroplasmy was present, with 50.92% of mutant mtDNA present.

MITOCHONDRIAL SPINOCEREBELLAR ATAXIA AND EPILEPSY

The small and maternally inherited genome found inside mitochondria encodes 13 subunits of the respiratory chain. DNA polymerase γ (pol γ) is the enzyme that replicates and repairs the mtDNA. Pol γ is a heterotrimer composed of one catalytic subunit (pol γA), and two accessory subunits (pol γB). Over 120 pathogenic mutations have been described in the gene encoding the catalytic pol γ subunit (POLG), and these are associated with a wide spectrum of neurological syndromes, ranging from adult onset myopathies to severe infantile encephalopathies. MSCAE is inherited as a recessive disorder most commonly associated with the mutations c.1399G>A, p.A467T or/and c.2243G>C, p.W748S in the linker region of pol γA.

PolγA comprises a polymerase (replicating) domain and an exonuclease (proofreading) domain, separated by a large linker region. The linker domain is the binding site of the accessory subunits, which enhance substrate affinity and processivity of the catalytic subunit. The A467T interferes with the catalytic subunit's intrinsic polymerase activity and binding to the accessory subunit, resulting in severely reduced efficiency of mtDNA synthesis. It is possible the W748S mutation has a similar effect. All these pol γ mutations induce mtDNA damage in the form of point mutations, multiple deletions, and quantitative depletion.

While ataxia and peripheral neuropathy are predominant features present in the 100% and 98% of the patients with *MSCAE*, respectively, headache is observed in 83% and epilepsy in 68% of them. A variety of clinical seizure types are seen, including partial simple or complex visual and motor seizures and GTC seizures. Most common are simple partial motor seizures involving an upper limb and the head/neck region, and these often evolve into epilepsia partialis continua (EPC), which may last for up to several months. Visual seizures are common, and patients usually describe flashing colored lights in one or both visual hemifields. Patients with epilepsy experience episodes of clinical exacerbation with severe seizures and rapidly progressing encephalopathy. The course of MSCAE is invariably progressive.

The rate of progression and mortality are highly variable and linked to two factors: genotype and epilepsy. Survival is worse in patients carrying the A467T and W748S mutations (compound heterozygous) and best in A467T homozygotes [80].

ALPERS' DISEASE

Alpers' disease is a childhood-onset epileptic syndrome produced by a mutation in nuclear polymerase γ (*POLG-A*) gene and characterized by hepatocerebral disorder.

All these observations confirm that deficiencies in mitochondrial energy production can induce seizures and raise the interesting question of whether mutations in mtDNA as well as DNA polymerase γ could contribute to the unexplained but well-documented maternal influence on the transmission of epilepsy.

Leucine-rich Glioma Inactivated Gene 1 (*LGI1*)

Autosomal dominant lateral temporal lobe epilepsy (ADLTE, also known as autosomal dominant partial epilepsy with auditory features) is an idiopathic syndrome that is characterized by simple partial seizures with mainly acoustic and sometimes even visual hallucinations. It was first described in a three-generation family with 11 members diagnosed with an idiopathic/cryptogenic epilepsy [81]. In some families, the seizures can start with a brief sensory aphasia without reduced consciousness [82]. Because all the genes that were previously implicated in idiopathic epilepsy coded for ion channel subunits, the identification of leucine-rich glioma inactivated gene 1 (*LGI1*), located on chromosome 10q24, as the gene responsible for ADLTE came as a surprise. Although the function of the *LGI1* protein is still unknown, sequence analysis showed that it is not an ion channel subunit, and *LGI1* mutations have been found in ADLTE families [83, 84]. Similarly, an *in silico* study showed that *LGI1* is probably not membrane-bound and shares the formerly unknown epitempin repeat with another epilepsy gene, the *MASS1/VLGR1* gene. The epitempin repeat is located in the C-terminus and consists of a tandem repeat with a core of about 50 residues. The *MASS1/VLGR1* gene, which is mutated in the Frings mouse model for audiogenic seizures and in one family with febrile seizures, shares the epitempin repeat [85, 86].

The second hallmark of *LGI1*—the epitempin repeat—is located in the C-terminus and consists of a tandem repeat with a core of about 50 residues, which probably fold into a β-propeller structure 51. The function of the epitempin repeat is unknown, but it was also found in another gene that has been associated with epilepsy—the *MASS1/VLGR1* gene. *MASS1/VLGR1* codes for the very large G-protein-coupled receptor, and mutations in this gene are responsible for the Frings mouse model of audiogenic epilepsy 52. Mutations in the *MASS1/VLGR1* gene also seem to be a rare cause of febrile seizures in humans. In the *MASS1/VLGR1* protein, the epitempin repeat is part of the ligand-binding ectodomain; thus, like the LRR domain, this repeat might be involved in protein–protein interactions. This raises the question of whether this formerly unknown sequence signature might be involved in a new mechanism of epileptogenesis. One possibility is that the epitempin repeat might be important for mechanisms like synaptogenesis or axon guidance, which depend on the communication between cells or between cells and extracellular matrix proteins. In such a model, the epilepsy that is caused by mutations in the *LGI1* gene could be the result of abnormalities in synapse formation or neuronal migration.

LGI1 was first cloned from the translocation breakpoint of a glioma cell line and was subsequently shown to be downregulated in many, but not all, glioblastoma cell lines that were tested. As no mutations or deletions of *LGI1* were found in these cell lines, it remains unclear whether the loss of *LGI1* expression is indeed a causative event in tumorigenesis. However, the re-expression of *LGI1* in cell lines that lack *LGI1* mRNA significantly reduced cell proliferation, supporting the hypothesis that *LGI1* is a tumor suppressor gene. No evidence has been found that the *LGI1* mutations in families with ADLTE increase the rate of brain tumors or other malignancies, excluding a role for *LGI1* as a first-step or high-penetrance tumor suppressor gene.

Inborn Errors of Metabolism and Epilepsy

Inborn errors of metabolism (IEMs) are a group of genetic disorders characterized by the dysfunction of an enzyme or other protein involved in cellular metabolism. To date, more than 11,000 inherited disorders in humans have been described, including around 500 enzymatic deficits affecting the metabolism of intracellular organelles, energy metabolism, amino acid catabolism, synthesis and degradation of proteins and lipids, etc.

Almost 200 of these are associated with seizures and epilepsy. IEMs are a group of genetic disorders characterized by dysfunction of an enzyme or other protein involved in cellular metabolism. Most IEMs involve the nervous system (neurometabolic diseases, or NMD). NMDs often present with a complex clinical picture: psychomotor retardation and/or regression, pyramidal signs, ataxia, hypotonia, movement disorders, and epilepsy. Furthermore, epilepsy is more frequently part of this complex picture than a predominant symptom and presents with various clinical and EEG features. Various NMD-causing epilepsy may be suspected according to age at onset and the presence or absence of specific epileptic syndromes. A wide spectrum of clinical presentation of the seizures ranging from generalized seizures (tonic–clonic, absence, clonic, tonic, atonic, myoclonus), partial seizures (partial simple and partial complexes), and particular syndromes as West syndrome [87].

Epilepsy and NMDs

The development of epilepsy during the course of a NMD can affect almost half of the patients. Seizures are a common symptom in a great number of metabolic disorders, occurring mainly in infancy and childhood. In some, seizures may occur only until adequate treatment is initiated or as a consequence of acute metabolic decompensation, as is the case in hyperammonemia or hypoglycemia. In others, seizures can be the main manifestation of the disease and can lead to AED-resistant epilepsy, e.g., in one of the creatine deficiency syndromes, guanidinoacetate methyltransferase (GAMT) deficiency.

In the context of an epileptic syndrome, the presence of NMD should be suspected when the patient presents one of several features such as

- medical history shows consanguinity, family history of similar cases, and onset at the neonatal period or psychomotor delay
- potential association with a neurological deterioration or systemic signs
- seizures worsen under certain AEDs such as valproate
- presence of myoclonus, spasms, focal and tonic seizures, unexplained status epilepticus, or seizures related to eating times
- early and/or progressive myoclonic epilepsy (PME)

Most hereditary NMD belong to the single-organelle multiorgan disorders, which include lysosomal, peroxisomal, and mitochondrial conditions, of which lysosomal diseases are the best explored ones.

Various NMD-causing epilepsies may be suspected according to age at onset and the presence or absence of specific epileptic syndromes. These diseases can be classified according to the nature of their physiopathology (Table 7A), according to age at onset (Table 7B), or according to the type of presenting seizures or epilepsy syndrome (Table 7C) as well as the epilepsies amenable to metabolic treatment (Table 7D) [88].

During the neonatal period, early myoclonic encephalopathy (EME) with suppression-burst EEG pattern is often due to an IEM, particularly nonketotic hyperglycinemia. Furthermore, if seizures are unexplained and refractory, vitamin-responsive epilepsies should always be considered

TABLE 7
A. According to the nature of their physiopathology
• Energy deficiency: hypoglycemia, *GLUT1* deficiency, respiratory chain deficiency, creatine deficiency
• Toxic effect: amino acidopathies, organic acidurias, urea cycle defects
• Impaired neuronal function: storage disorders
• Disturbance of neurotransmitter systems: nonketotic hyperglycinemia, GABA transaminase deficiency, succinic semialdehyde dehydrogenase deficiency
• Associated brain malformations: peroxisomal disorders (Zellweger), O-glycosylation defects
• Vitamin/cofactor dependency: biotinidase deficiency, pyridoxine-dependent, and pyridoxal phosphate-dependent epilepsy, folinic acid-responsive seizures, Menkes' disease
• Miscellaneous: congenital disorders of glycosylation, serine biosynthesis deficiency, and inborn errors of brain excitability (ion channel disorders)
B. According to age at onset
• Neonatal period: hypoglycemia, pyridoxine dependency, PNPO deficiency, nonketotic hyperglycinemia, organic acidurias, urea cycle defects, neonatal adrenoleukodystrophy, Zellweger syndrome, folinic acid-responsive seizures, holocarboxylase synthase deficiency, molybdenum cofactor deficiency, sulfite oxidase deficiency
• Infancy: hypoglycemia, *GLUT1* deficiency, creatine deficiency, biotinidase deficiency, amino acidopathies, organic acidurias, congenital disorders of glycosylation, pyridoxine dependency, infantile form of NCL (NCL1)
• Toddlers: late infantile form of NCL (NCL2), mitochondrial disorders including Alpers' disease, lysosomal storage disorders
• School age: mitochondrial disorders, juvenile form of NCL (NCL3), progressive myoclonus epilepsies

TABLE 7 (*continued*)
C. *According to the type of presenting seizures or epilepsy syndrome*
• Infantile spasms: biotinidase deficiency, Menkes' disease, mitochondrial disorders, organic acidurias, amino acidopathies
• Epilepsy with myoclonic seizures: nonketotic hyperglycinemia, mitochondrial disorders, *GLUT1* deficiency, storage disorders
• Progressive myoclonic epilepsies: Lafora disease, MERRF, MELAS, Unverricht–Lundborg disease, sialidosis
• Epilepsy with GTC seizures: *GLUT1* deficiency, NCL2, NCL3, other storage disorders, mitochondrial disorders
• Epilepsy with myoclonic–astatic: seizures *GLUT1* deficiency, NCL2
• Epilepsy with (multi-)focal seizures: NCL3, *GLUT1* deficiency, and others
• EPC: Alpers' disease, other mitochondrial disorders
D. *Epilepsies amenable to metabolic treatment*
• *GLUT1* deficiency: ketogenic diet
• Cofactor-dependent epilepsy: pyridoxine, pyridoxal phosphate, folinic acid, biotin
• GAMT deficiency: creatine supplementation, arginine-restricted, ornithine-enriched diet
• Phenylketonuria: low-phenylalanine diet; in atypical phenylketonuria, substitution of L-DOPA, 5OH-tryptophan, folinic acid
• Defects of serine biosynthesis: serine supplementation

and therapeutic trials using successively pyridoxine, pyridoxal phosphate, folinic acid, and biotin should be started. In early infancy, many treatable conditions should be considered: serine deficiency, *GLUT1* deficiency, biotinidase deficiency, aminoacidopathies, and urea cycle defects.

Additionally, other less readily treatable diseases (mitochondriopathies, sulfite oxidase deficiency, adenylosuccinate lyase deficiency, succinic semialdehyde dehydrogenase, peroxysomal disorders, and CDG syndromes) could also be searched. In childhood (as it is the case in adulthood), diagnosis approach is based on the presence or absence of PME, including ceroide-lipofuscinosis, sialidosis, GM2 gangliosidosis, Gaucher disease type III, Niemann–Pick type C, Lafora disease, and mitochondriopathies.In the absence of PME, treatable diseases (such as *GLUT1* deficiency, creatine deficiency, porphyria, and urea cycle disorders) should be considered. Interestingly, epilepsy in NMD is usually pharmacoresistant, and some AEDs (such as valproate) may worsen the clinical condition as they interfere with the abnormal metabolic pathway, especially in mitochondriopathies, urea cycle deficiency, and nonketotic hyperglycinemia [89].

Disease	Gene	Protein	Chromosomal Locus
Maple syrup disease	BCKDHA	branched chain ketoacid dehydrogenase $E_1\alpha$ polypeptide	19q13.1-q13.2
	BCKDHB	branched chain ketoacid dehydrogenase $E_1\beta$ polypeptide	6q14.1
	DBT	dihydrolipoamide branched chain transacylase E2	1p31
	DLD	dihydrolipoamide dehydrogenase	7q31-q32
Glutaric acidemia type II	ETFA	electron-transfer-flavoprotein, alpha polypeptide	15q23-q25
	ETFB	electron-transfer-flavoprotein, beta polypeptide	19q13.3
	ETFDH	electron-transferring-flavoprotein dehydrogenase	4q32-q35
Methylmalonic acidemia	MUT	methylmalonyl CoA mutase	6p12.3
	MMAA	methylmalonic aciduria (cobalamin deficiency) cblA type	4q31.21
	MMAB	methylmalonic aciduria (cobalamin deficiency) cblB type	12q24
	MADHC	methylmalonic aciduria (cobalamin deficiency) cblD	2q23.2
	MCEE	methylmalonyl CoA epimerase	2p13.3
Propionic acidemia	PCCA	propionyl CoA carboxylase, alpha polypeptide	13q32
	PCCB	propionyl CoA carboxylase, beta polypeptide	3q21-q22
Isovaleric acidemia	IVD	isovaleryl-CoA dehydrogenase	15q14-q15
β-oxidation of fatty acids	CPT2	carnitine palmitoyltransferase II	1p32
	HADHA	long-chain 3-hydroxyacyl-CoA dehydrogenase (LCHAD)	2p23
	HADHA&B	mitochondrial trifunctional protein	2p23
Pyruvate metabolism		Pyruvate dehydrogenase complex	
	X-linked E1α	• Pyruvate dehydrogenase (E1)	Xp22.12
	DLAT	• Dihydrolipoyl transacetylase (E2)	11q23.1
	DLD	• Dihydrolipoyl dehydrogenase (E3)	7q31.1
	PC	Pyruvate carboxylase	11q13.4-q13.5

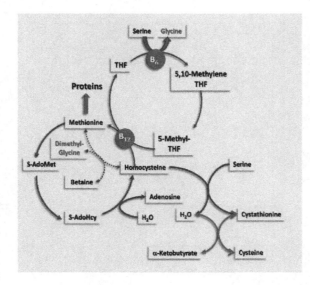

FOLR1 Gene Mutation

Several unrelated disorders can lead to 5-methyltetrahydrofolate (5MTHF) depletion in the cerobrospinal fluid (CSF), including primary genetic disorders in folate-related pathways or those causing defective transport across the blood–CSF barrier. A case of cerebral folate transport deficiency due to a novel homozygous mutation in the *FOLR1* gene has been reported. A previously healthy infant developed an ataxic syndrome in the second year of life, followed by choreic movements and PME. At the age of 26 months, measurement of 5-MTHF in CSF by HPLC with fluorescence detection and conducted magnetic resonance (MR) imaging and spectroscopy studies were developed. A mutational screening in the coding region of the *FOLR1* gene was done. MRI showed a diffuse abnormal signal of the cerebral white matter, cerebellar atrophy, and a reduced peak of choline in spectroscopy. A profound deficiency of CSF 5MTHF (2 nmol/L; NV 48–127) with reduced CSF/plasma folate ratio (0.4; NV 1.5–3.5) was highly suggestive of defective brain folate-specific transport across the blood–CSF/brain barrier (BCSFB and BBB, respectively). Mutation screening of *FOLR1* revealed a new homozygous missense mutation (p.Cys15Arg) that is predicted to abolish a disulfide bond, probably necessary for the correct folding of the protein. Both parents were heterozygous carriers of the same variant. Mutation screening in the *FOLR1* gene is advisable in children with profound 5MTHF deficiency and decreased CSF/serum folate ratio. Progressive ataxia and myoclonic epilepsy, together with impaired brain myelination, are clinical hallmarks of the disease (Figure 17).

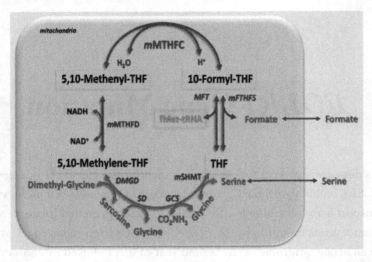

FIGURE 17: Compartmentation of folate-mediated one-carbon metabolism in mitochondria. One-carbon metabolism in mitochondria is required to generate formate for one-carbon metabolism in the cytoplasm, to generate the amino acid glycine, and to synthesize formylmethionyl tRNA for protein synthesis in mitochondria. DMGD, dimethylglycine dehydrogenase; GSC, glycine cleavage system; mFTHFS, mitochondrial formyltetrahydrofolate synthetase; mMTHFC, mitochondrial methenyltetrahydrofolate cyclohydrolase; mMTHFD, mitochondrial methylenetetrahydrofolate dehydrogenase; mSHMT, mitochondrial serine hydroxymethyltransferase; M-tRNA-FT, methionyl-tRNA formyltransferase; MFT, mitochondrial folate transporter; SD, sarcosine dehydrogenase.

Lysosomal Storage Disorders

There are severe genetic diseases caused by the defective activity of lysosomal proteins, cofactors, or integral membrane proteins, which result in the intralysosomal accumulation of undegraded metabolites such as sphingolipids, cholesterol, glycoproteins, mucopolysaccharides, or glycogen, and together, the combined frequency of lysosomal storage disorders (LSDs) is estimated to be approximately 1 in 8000 live births. More than 50 LSDs have been described to date [90]. Although they are characterized by a wide spectrum of clinical phenotypes, many of these disorders present with severe progressive neurological impairment. Among the neurological symptoms, the presence of PME has been reported in different LSDs, including Gaucher disease, action myoclonus–renal failure syndrome, NCL, sialidosis, Niemann–Pick type C disease, and GM_2 gangliosidosis.

Neuronal Ceroid Lipofuscinoses (NCL)				
Gene	Chromosome (n° nutations)		Protein	Main storage material
CNL1	1p32	48	palmitoyl proteinthioesterase (PPT1), lysosomal enzyme	Saposins A and D
CNL2	11p12	72	tripeptidil peptidase 1 (TPP1), lysosomal enzyme	SC-ATPs*
CNL3	16p12	49	CNL3, lysosomal transmembrane protein	SC-ATPs
CNL5	13q21-q32	27	CNL5, lysosomal soluble protein	SC-ATPs
CNL6	15q23	46	CNL6, transmembrane ER protein	SC-ATPs
CNL7	4q28.1-q28.2	23	CNL7, lysosomal transmembrane protein	SC-ATPs
CNL8	8p23	16	CNL8, transmembrane ER protein	SC-ATPs
CNL10	11p15.5	4	CTSD, cathepsin D, lysosomal enzyme	SC-ATPs

•SC-ATPs (Subunit -C of ATP synthase)

Irrespective of major general clinical presentations with or without seizures, some lysosomal diseases share similar phenotype features as the presence of the "cherry-red spot." This finding led to the diagnosis of a few conditions such as sialidosis, Neimann–Pick (type B), Tay–Sachs disease, Sandhoff disease, and Farber disease.

LSD is caused by the inherited deficiency of the lysosomal enzyme α-N-acetyl-neuraminidase 1 (NEU1), which cleaves the terminal sialic acid residues of several oliogosaccharides and polypeptides. The deficiency of NEU1 leads to the accumulation of sialic acid (N-acetylneuraminic acid) covalently linked to oligosaccharides and/or glycoproteins. This aspect distinguishes sialidoses from sialurias, in which the neuraminidase activity is normal or elevated with a storage and excretion of "free" sialic acid rather than "bound" forms. Two main clinical variants have been described:

type I, the milder form of the disease, which lacks the physical changes (normosomatic) and may present seizures, and type II, a more severe form with an earlier onset, which can be subdivided in two different phenotypes: congenital/neonatal and juvenile forms. The neonatal form presents severe dysmorphisms (coarse feces, pectus carinatum, short trunk, exaggerated thoracic kyphosis, and wadding gait) as well as growth delay. Infantile phenotypes are characterized by cherry-red spot, corneal opacity, hearing loss, progressive neurodegeneration, and cognitive deterioration with myoclonic seizures, whereas the juvenile onset is characterized by less pronounced dysmorphic signs with muscular hypotonia and hypotrophy, ataxia, and myoclonic seizures.

The Solute-carrier Gene Superfamily and Epilepsy

The solute-carrier gene (*SLC*) superfamily encodes membrane-bound transporters and comprises 55 gene families having at least 362 putatively functional protein-coding genes.

Gene products include passive transporters, symporters, and antiporters located in all cellular and organelle membranes except, perhaps, the nuclear membrane. Transport substrates include amino acids and oligopeptides, glucose and other sugars, inorganic cations and anions (H^+, HCO_3^{2-}, Cl^-, Na^+, K^+, Ca^{2+}, Mg^{2+}, PO_4^{3-}, HPO_4^{2-}, $H_2PO_4^-$, SO_4^{2-}, $C_2O_4^{2-}$, OH^-, CO_3^{2-}), bile salts, carboxylate and other organic anions, acetyl coenzyme A, essential metals, biogenic amines, neurotransmitters, vitamins, fatty acids and lipids, nucleosides, ammonium, choline, thyroid hormone, and urea. However, despite the wide spectrum of solutes transported by this SCL superfamily, few mutations on these transporters are associated with epileptic syndromes.

(a) Defects in SLC25A22 (a mitochondrial glutamate carrier) are the cause of epileptic encephalopathy early infantile type 3 (EIEE3), also known as EME or neonatal epilepsy with suppression-burst pattern. Severe neonatal epilepsies with suppression-burst pattern are early-onset epileptic syndromes characterized by a typical EEG pattern. The suppression-burst pattern of the EEG is characterized by higher-voltage bursts of slow waves mixed with multifocal spikes, alternating with isoelectric suppression phases. EME is characterized by a very early onset, erratic, and fragmentary myoclonus, massive myoclonus, partial motor seizures, and late tonic spasms. The prognosis of EME is poor, with no effective treatment, and children with the condition either die within 1 to 2 years after birth or survive in a persistent vegetative state. EME inheritance is autosomal recessive.

(b) Christianson syndrome: Mutations in the *SLC9A6* gene, related with Na^+/H^+ exchanger isoform 6, lead to this condition, characterized by neurological problems, including intellectual disabilities, seizures, and an inability to walk or speak. Mutations in the *SLC9A6* gene typically lead to an abnormally short NHE6 protein that is nonfunctional or that is broken down quickly in cells, resulting in the absence of functional NHE6 channels. As a result, the pH in endosomes is not properly maintained. It is unclear how unregulated endosomal pH leads to neurological problems in people with Christianson syndrome. Some studies have shown that protein trafficking by endosomes is important for learning and memory, but the role of endosomal pH or the NHE6 protein in this process has not been identified. Recently, it was described that Na^+/H^+ exchanger isoform 6 (*NHE6/SLC9A6*) is involved in clathrin-dependent endocytosis of transferring [91]. This finding suggests that an iron-dependent neurodegenerative process could be present in this epileptic syndrome.

(c) X-linked creatine deficiency: caused by at least 20 mutations mutations in the *SLC6A8* gene have been identified in people with X-linked creatine deficiency, a disorder that causes intellectual disability, behavioral problems, seizures, and muscle weakness. *SLC6A8* gene mutations impair the ability of the transporter protein to bring creatine into cells, resulting in a creatine shortage (deficiency). The effects of creatine deficiency are most severe in organs and tissues that require large amounts of energy, especially the brain.

(d) Cystinuria: Some people with cystinuria have large DNA deletions that remove not only the *SLC3A1* gene but one or more neighboring genes. Individuals with these large DNA deletions have the signs and symptoms of cystinuria, but they can also have other features. Deletions of the *SLC3A1* gene and the neighboring *PREPL* gene cause hypotonia–cystinuria syndrome. In addition to cystinuria, people with this condition have low muscle tone (hypotonia) and poor feeding, which usually improves by early childhood. They may also have droopy eyelids (ptosis), an elongated head (dolichocephaly), and mild intellectual disability. Most people with this condition have short stature. Few cases of cystinuria associated with seizures or epilepsy were reported [91]. Deletions of the *SLC3A1* gene, the *PREPL* gene, the *C2orf34* gene, and the *PPM1B* gene cause 2p21 deletion syndrome. In addition to all the symptoms of the previous syndromes, individuals with 2p21 deletion syndrome have seizures soon after birth, moderate to severe psychomotor delay, and impairments in the process from which cells derive much of their energy (oxidative phosphorylation). People with this condition typically have a characteristic facial appearance with a prominent forehead, long eyelashes, a flat nasal bridge, and abnormally turned ears.

(e) BBB glucose transport defect: Defects in SLC2A1 are the cause of *GLUT1* deficiency syndrome type 1 (GLUT1DS1), also known as BBB glucose transport defect. It is a neurological disorder showing wide phenotypic variability. The most severe "classic" phenotype comprises infantile-onset epileptic encephalopathy associated with delayed development, acquired microcephaly, motor incoordination, and spasticity. Onset of seizures, usually characterized by apneic episodes, staring spells, and episodic eye movements, occurs within the first 4 months of life. Other paroxysmal findings include intermittent ataxia, confusion, lethargy, sleep disturbance, and headache. Varying degrees of cognitive impairment can occur, ranging from learning disabilities to severe mental retardation.

Genes Related with Different Epileptic Syndromes

Generalized Epilepsies (36 genes)
Myoclonic Epilepsy , Febrile Seizures, Absences

Gene	Protein altered	Chromosome locus
ALDH7A1	Aldehyde dehydrogenase 7 family, member A1	5q31
BRD2	Bromodomain-containing protein 2	6p21.3
CACNA1A	Ca channel, voltage-dependent P/Q type, α 1A subunit	19p13
CACNA1H	Ca channel, voltage-dependent T type α 1H subunit	16p13.3
CACNB4	Ca channel, voltage-dependent β 4 subunit	2q22-q23
CASR	Calcium-sensing receptor	3q21.1
CHRNA2	Cholinergic receptor, nicotinic α2 (neuronal)	8p21
CHRNA4	Cholinergic receptor, nicotinic α 4 (neuronal)	20q13.2-q13.3
CHRNB2	Cholinergic receptor, nicotinic β2 (neuronal)	1q21.3
CLCN2	Chloride channel, voltage-sensitive 2	3q27-q28
CSTB	Cystatin B (stefin B) ↑ dodecamer repeat CCCCG-CCCCG-CG	21q22.3
EFHC1	EF-hand domain (C-terminal) containing 1	6p12.3
EPM2A	Laforin (PME type 2A, Lafora disease)	6q24
GABRA1	Gamma-aminobutyric acid (GABA) A receptor, α 1	5q34
GABRB3	Gamma-aminobutyric acid (GABA) A receptor β3	15q12
GABRD	Gamma-aminobutyric acid (GABA) A receptor, δ	1p|1p36.3
GABRG2	Gamma-aminobutyric acid (GABA) A receptor, γ 2	5q34
GPR98	G protein-coupled receptor 98	5q13
GRIN2A	Glutamate receptor, ionotropic, N-methyl D-aspartate 2A	16p13.2
GRIN2B	Glutamate receptor, ionotropic, N-methyl D-aspartate 2B	12p1
KCNMA1	K large conductance Ca-activated channel, subfamily M, α1	10q22
KCNQ2	K voltage-gated channel, KQT-like subfamily, member 2	20q13.33
KCNQ3	K voltage-gated channel, KQT-like subfamily, member 3	8q24
KCTD7	K channel tetramerisation domain containing 7	7q11.21
ME2	Malic enzyme 2, NAD(+)-dependent, mitochondrial	18q21
NHLRC1	NHL repeat containing 1	6p22.3
PCDH19	Protocadherin 19	Xq22.1
PRICKLE1	Prickle homolog 1 (Drosophila)	12p11-q12

PRICKLE2	Prickle homolog 2 (Drosophila)	3p14.3
SCARB2	scavenger receptor class B, member 2	4q21.1
SCN1A	Na channel, voltage-gated, type I,α subunit	2q24.3
SCN2A	Na channel, voltage-gated, type II,α subunit	2q24.3
SCN9A	Na channel, voltage-gated, type IX, α subunit	2q24
SCN1B	Na channel, voltage-gated, type I,β subunit	19q13.1
SLC2A1	Solute carrier family 2 (facilitated glucose transp) 1	1p34.2
TBC1D24	TBC1 domain family, member 24	16p13.3

Epilepsy and X-linked Mental Retardation (25 genes)

Epileptic syndrome	Gene	Protein altered	Chromosome locus
Early Infantile Epileptic Encephalopathy	ARHGEF9	Cdc42 guanine nucleotide exchange factor-9	Xq11.1
Early Infantile Epileptic Encephalopathy	ARX	Aristaless related homeobox	Xp21.3
Epilepsy with XLMR	ATP6AP2	ATPase, H⁺ transp, lysosomal accessory prot. 2	Xp11.4
Epilepsy with XLMR	ATRX	α thalassemia/mental retard. Synd. X-linked	Xq21.1
Mental Retardation and Microcephaly	CASK	Ca/calmodulin-dependent SPK	Xp11.4
Early Infantile Epileptic Encephalopathy	CDKL5	Cyclin-dependent kinase-like 5	Xp22
Epilepsy with XLMR	CUL4B	Cullin 4B	Xq23
Simpson-Golabi-Behmel Syndrome	CXORF5	OFD1	Xp22.3-22.2
Lissencephaly	DCX	Doublecortin	Xq22.3-q23
Aarskog Scott Syndrome	FGD1	FYVE, RhoGEF and PH domain containing 1	Xp11.21
Simpson-Golabi-Behmel Syndrome	GPC3	Glypican 3	Xq26.1
Epilepsy with XLMR	GRIA3	Glutamate receptor, ionotropic, AMPA 3	Xq25
Epilepsy with XLMR	HSD17B10	Hydroxysteroid-17–β-dehydrogenase-10	Xp11.2
Epilepsy with XLMR	JARID1C	Lysine (K)-specific demethylase 5C	Xp11.22-p11.21
Epilepsy with XLMR	OPHN1	Oligophrenin 1	Xq12
Epilepsy with XLMR	PAK3	p21 protein (Cdc42/Rac)-activated kinase 3	Xq23
Boerjeson Forsmann Lehmann Syndrome	PHF6	PHD finger protein 6	Xq26.3
Pelizaeus-Merzbacher Disease	PLP1	Proteolipid protein 1	Xq22
Epilepsy with XLMR	PQBP1	Polyglutamine binding protein 1	Xp11.23
Epilepsy with XLMR	RAB39B	RAB39B, member RAS oncogene family	Xq28
Angelman-Like Syndrome	SLC9A6	Solute carrier family 9, subfamily A-member 6	Xq26.3
Cornelia De Lange Syndrome	SMC1A	Structural maintenance of chromosomes 1A	Xp11.22-p11.21
Epilepsy with XLMR	SMS	Spermine synthase	Xp22.1
Rolandic Epilepsy	SRPX2	Sushi-repeat containing protein, X-linked 2	Xq21.33-q23
Epilepsy with XLMR	SYP	Synaptophysin	Xp11.23-p11.22

Epileptic Encephalopathies (30 genes)

Epileptic syndrome	Gene	Protein altered	Chromosome locus
Aicardi-Goutieres Syndrome	RNASEH2A	Ribonuclease H2, large subunit	19p13.2
	RNASEH2B	Ribonuclease H2, subunit B	13q14.3
	RNASEH2C	ribonuclease H2, subunit C	11q13.1
	SAMHD1	SAM domain and HD domain 1	20pter-q12
	TREX1	3- prime repair exonuclease 1	3p21.31
Angelman Syndrome	SLC9A6	Solute carrier family 9, subfamily A-6	Xq26.3
	UBE3A	Ubiquitin protein ligase E3A	15q11.2
GLUT1 Deficiency Syndrome	SLC2A1	Solute carrier family 2 , member 1	1p34.2
Early Infantile Epil Encephalop	ARHGEF9	Cdc42 guanine nucleotide exchange factor-9	Xq11.1
	ARX	Aristaless related homeobox	Xp21.3
	CDKL5	Cyclin-dependent kinase-like 5	Xp22
	GABRG2	GABA-A receptor-γ2	5q34
	GRIN2A	Glutamate receptor, ionotropic, NMDA-2A	16p13.2
	GRIN2B	Glutamate receptor, ionotropic, NMDA-2B	12p12
	PCDH19	Protocadherin 19	Xq22.1
	PNKP	Polynucleotide kinase 3'-phosphatase	19q13.3-q13.4
	SCN1A	Na channel, voltage-gated, type I, α subunit	2q24.3
	SCN1B	Na channel, voltage-gated, type I, β subunit	19q13.1
	SCN2A	Na channel, voltage-gated, type II, α subunit	2q24.3
	SCN9A	Na channel, voltage-gated, type IX, α subunit	2q24
	SLC25A22	Mitochondrial glutamate carrier member 22	11p15.5
	SPTAN1	Spectrin, alpha, non-erythrocytic 1	9q34.11
	STXBP1	Syntaxin binding protein 1	9q34.1
Lennox Gastaut Syndrome	MAPK10	Mitogen-activated protein kinase 10	4q22.1-q23
Mowat-Wilson Syndrome	ZEB2	Zinc finger E-box binding homeobox 2	2q22.3
Pitt Hopkins Syndrome	TCF4	Transcription factor 4	18q21.1
	CNTNAP2	Contactin associated protein-like 2	7q35
	NRXN1	Neurexin 1	2p16.3
Rett Syndrome	FOXG1	Forkhead box G1	14q13
	MECP2	Methyl CpG binding protein 2	Xq28

Genetic Mechanisms of Drug Resistance in Epilepsy

Epilepsies are among the most common neurological disorders, affecting up to 1% of the population. Most patients with epilepsy become seizure-free with AED therapy. However, approximately 30%–40% of epilepsy patients have a medically intractable condition, even if they are treated with a combination of various AEDs at maximal dosages. Several questions arise from this clinical evidence, and there is a need to find answers about the genetic–molecular bases of intrinsic and/or acquired factors related to this therapeutic behavior and perhaps also related to the severity of the disease. A few hypotheses based on clinical and experimental evidence have been proposed to get a better understanding of this phenomenon.

Definition of Drug Resistance in Epilepsy

Resistance to drug treatment is a critical problem in the therapy of many brain disorders including epilepsy. Observational cohort studies of newly diagnosed epilepsy in adults and children suggest that once a patient has failed trials of two appropriate AEDs, the probability of achieving seizure freedom with subsequent AED treatments is modest. How prevalent is drug resistance in epilepsy, and what is the correct definition of RE medication? Although patients with drug resistance can be easily recognized in current clinical practice, perhaps, the main starting point must be the definition of the drug-resistant phenotype.

In the daily clinical follow-up of epileptic patients, the failure to control seizures despite use of two or more appropriate AEDs, even when maximum tolerated doses are administered, is a useful functional criteria of refractoriness. This multidrug-resistance phenotype (Figure 18b) may be acquired during the progress of the epileptic syndrome, and could also be present in the early stage of the disease [92]. Why does a subgroup of patients repeatedly fail to control the seizures with one AED after another (Figure 19a)? During the last few decades, more than 15 new AEDs have become available; however, the percentage of patients with RE remains near 30%–40%, as observed during the early era of treatment with bromide [93] (Figure 19b).

In the pediatric epileptic population, upon first treatment, patient groups are divide into early remission (31%), definitive nonresponders (19%), and a third group, which is considered as potential late remission (50%). However, both remission groups will change secondary to failures in the medication schedules, needing new or more drugs, and reaching a final situation with 67% of responders versus 33% with RE. The variability of the response to similar medications, conduce to trial and error until to get the definitive categorization of the patient [94] (Figure 20).

Sanjay M. Sisodiya discusses the genetics of drug resistance in epilepsy and indicates that patients who were considered drug resistant under a given definition may not remain so as newer AEDs are developed, or designed, to target previously unappreciated underlying pathophysiological mechanisms. Furthermore, he asks whether a patient with temporal lobe epilepsy due to hippocampal sclerosis fails to respond to carbamazepine, phenytoin (PHT), lamotrigine, and BZDs but

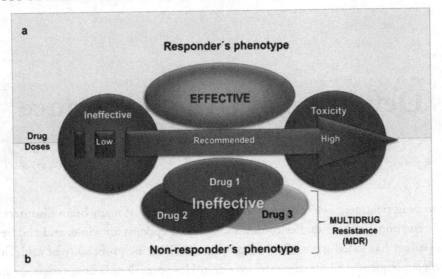

FIGURE 18: (a) A responder patient receiving recommended doses of an correctly selected AED will have an effective therapy with control of crises. (b) Refractoriness is observed in nonresponders, despite receiving recommended doses of several combined AEDs, showing a MDR phenotype.

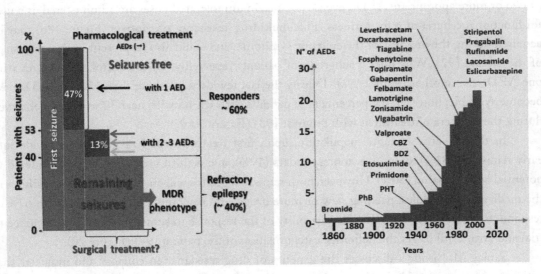

FIGURE 19: (a) After the administration of corresponding AEDs according with ILAE classification of epileptic syndromes after the first seizures, ~60%–70% of the patients will be seizure free. Remaining cases will develop RE with MDR phenotype. (b) New AEDs developed during last 20 years cannot modify the historic percentage of refractoriness.

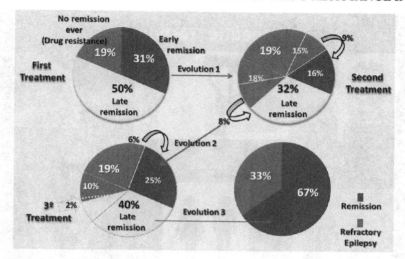

FIGURE 20: Adapted from "Natural study of treated childhood-onset epilepsy: Prospective long-term population-based study" by Silampaa and Schmidt Barin (2006).

becomes seizure free on levetiracetam or pregabalin would be considered drug resistant or not? Additionally, he suggests that for individuals who we could define as being drug resistant, perhaps his epilepsy is considered "drug resistant" simply because we do not yet have drugs that are appropriate for the treatment of that individual's epilepsy [95].

A main objective of pharmacotherapy is to get a positive result in the dose–response balance with "reasonable control" of the disease progress. However, sometimes, it is not achieved. As a response to the loss of efficacy observed in several common, a dose modification, new method of drug administration, or new drugs should always be considered. Recent advances in molecular medicine have led to the perspective that interindividual differences in responsiveness to drugs of various pharmacological classes, like many other human traits, are the result of a complex combination of genetic and environmental factors.

The study of individual genes whose variations exert a measurable influence on the effect of a given drug constitutes the field of *pharmacogenetics*, an area of medical sciences that has spawned a new age of inquiry into the relationship between gene structure/function and the effects of drugs.

Pharmacogenetics and pharmacogenomics could explain all these modifications, and give us the bases to use new biological markers and/or functional studies as potential predictive parameters to help choose the best treatments, know their potential efficacy, and avoid refractoriness (Figure 21).

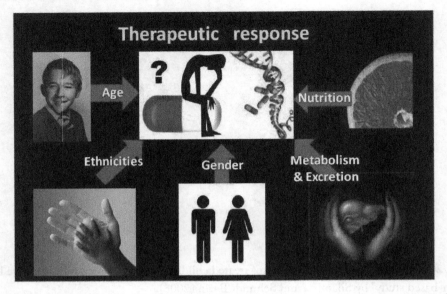

FIGURE 21: Several factors can alter the expected response to therapeutics: the constitutive differences observed in ethnic and gender groups, modifications occurring with the age, nutritional habits, different structural organs, and/or metabolic conditions. In all cases, there us underlying genetic predisposition that can increase or reduce phenotypic differences according to environmental influence.

After drugs gain access to the body, drugs can suffer several modifications to their structure and to their concentrations in circulation and tissues, which can modify the spectrum of their pharmacological effects as well as their intensity and duration. Thus, the extent to which a drug can be absorbed and transported to their targets influences drug potency and their effectiveness profile. The body's ability to metabolize and eliminate the drug and its metabolites affects the magnitude and duration of drug effects. These pharmacokinetic processes can change the outcome of drug therapy.

The scientific prerequisites for developing extended-release formulations of AEDs are that fluctuations in serum drug levels should reflect comparable fluctuations at the site of action (*brain*) and that minimizing such fluctuations is expected to be clinically beneficial. One complication of this situation that still remains to be resolved is that AEDs should act only on epileptic neurons of epileptogenic areas, mainly in focal epilepsies, but not on the rest of the brain. In these cases, the epileptogenic focus could need higher drug concentrations than the rest of the brain, and perhaps, the mechanisms of refractoriness are producing a paradoxical absurd situation, with reduced drug concentration at the focus but high drug exposure for normal neurons.

Genetics of Drug Response in Epilepsy

The genetic studies on drug response may prove to be more amenable to analysis than other aspects of genetics in epilepsy because the proteins that are drug targets, drug transporters, and drug metabolizers are, to varying extents, already known. It is well established that individuals who possess certain alleles of the *CYP2C9* gene, which encodes the major metabolizing enzyme of PHT, have significantly reduced rates of metabolism of PHT, necessitating lower maintenance doses, although prospective genotyping is not yet undertaken in practice. Whether variants in the *ABCB1* gene, which encodes the broad-spectrum multidrug transporter P-glycoprotein (P-gp), influence resistance to AEDs or not remains a hotly debated point. Gene variants influencing the sensitivity of targets to AEDs are also being uncovered: for example, a splice site variation in the *SCN1A* gene, which encodes the cerebral neuronal target of many AEDs, has been associated with dosing of these drugs. Such pharmacogenetic advances may permit closer modeling of treatment to the individual patient [96].

Most of the data about the bioavailability of orally administered AEDs are obtained from healthy control individuals. However, there is wide interindividual variation in the dose of AEDs required to achieve target blood concentrations. These variations can be related to mechanisms of drug absorption, biodistribution, metabolism, and excretion (ADME). Before any consideration of the molecular mechanisms related to the ADME system that can produce interindivual differences, we should mention that gastrointestinal or systemic diseases with altered intestinal, hepatic, or renal functions or the consequences of gastrointestinal surgery may modify drug absorption, metabolism, and excretion and alter the effectiveness of standard AED doses. AED absorption from the gastrointestinal tract is influenced by the gastric and intestinal motility, physicochemical properties of the environment in the small intestine, and surface area available for absorption. Many AEDs are administered to epileptic patients in whom an underlying concomitant disease might affect either their absorption or overall bioavailability. This is particularly true for patients suffering from diseases of intestinal absorption and digestion whom also have epilepsy. Several genetic conditions could produce an unspecific impairment of adequate AEDs absorption that includes both acquired and genetic disorders: primary malabsorption syndromes, such as congenital diseases with selective defect of single functions (e.g., lactose intolerance, sucrose–isomaltose intolerance of

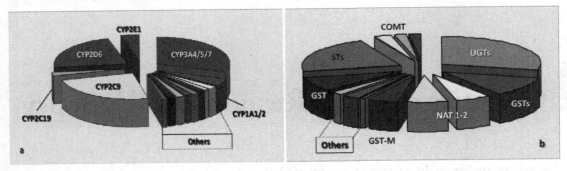

FIGURE 22: (a) Several isoforms of different enzymes (mainly CYP-P450 system) corresponding to the phase 1 of drug modification. (B) Isoforms corresponding to the drug conjugation of phase 2 are related with increased or decreased capacity to metabolize the drugs. Their functional expression can be repressed or induced by a wide spectrum of therapeutics, toxics, infections, inflammation, or by seizure stress itself. Furthermore, these modifications can be also influenced by the polymorphisms related with genetic differences as mentioned for the pharmacogenetics.

epithelial cells), or secondary malabsorption syndromes such as acquired small-intestinal diseases (celiac disease, tropical sprue, Whipple's disease, hypogammaglobulinemia, etc.).

The absorption of a drug and its distribution to various organs and tissues are processes governed not only by the physicochemical properties of the drug but also by endogenous molecules that interact with the drug.

Normally, routine monitoring of AED levels also demonstrate that drug concentration in serum fluctuates considerably during a dosing interval for those AEDs that are absorbed and eliminated rapidly. Genetic variation can potentially affect individual responsiveness to the drug at each of these steps. Whereas the field of pharmacokinetics addresses the way in which xenobiotics/drugs are affected by endogenous mechanisms, the field of pharmacodynamics refers to the actions, or molecular mechanisms by which xenobiotics/drugs produce their effects on the body, and it is here that individuals display variable drug effects despite equivalent drug concentrations. Genetic modifications on the pharmacological targets as well as on the proteins related with the drug ADME system are the bases of these different interindividual drug responses. The enzymes responsible for most of phase I metabolic reactions are mainly from the superfamily of enzymes CYP-450 (Figure 22a) and have an important impact on the biotransformation of drugs. CYPs are mainly expressed along the inner plasma membranes of mitochondria and the endoplasmic reticulum, and most CYP enzymes contribute, through oxidation, to the elimination of endogenous substrates and xenobiotics, enhancing their excretion from the body.

TABLE 8: Main enzymes of phase I that metabolize the common AEDs				
CYP2C8	*CYP2C9*	*CYP2C19*	*CYP3A4*	*CYP3A5*
Carbamazepine	Phenobarbital	Felbamate?	Carbamazepine	Carbamazepine
	Phenytoin	Clobazam	Clobazam	
	Primidone	Phenobarbital	Clonazepam	
	Valproic acid	Phenytoin	Ethosuximide	
		Valproic acid	Topiramate	
		Zonisamide	Zonisamide	

The CYP enzymes are encoded by 57 human genes whose products are involved in oxidative drug metabolism, as well as the synthesis of cholesterol, steroids, prostacyclins, and thromboxane. Most AEDs, except for gabapentin, lamotrigine, and levetiracetam, are metabolized at least partially by CYP enzymes (Table 8).

CYP2C9 and *CYP2C19* share >90% sequence similarity; however, differences within the substrate recognition site (SRS) regions, which are hypervariable areas, could be responsible for substrate selectivity (Figure 23). These regions include the phenylalanine cluster in the active site of *CYP2C9* composed of Phe100 and Phe114 (both B–C loop, SRS1) and Phe476 (b4-fold, SRS6) [97] (Table 9).

According to the number (i.e., 0, 1, and 2, respectively) of the defective alleles(s) of each *CYP2C* gene, individuals can be classified into three subgroups such as extensive, normal and poor drug metabolizers.

CYP2C9 and *CYP2C19* have well-characterized functional variants. In Caucasians, the frequencies of defective *CYP2C9*2* and *CYP2C9*3* are higher than in Asians (18.9% versus 2.5%–3.5%, respectively), and conversely, the frequencies of defective alleles *CYP2C19*2* and *CYP2C19*3* are higher in Asians than Caucasians (33%–43.5% versus 13.6%, respectively).

CYP3A metabolize carbamazepine, diazepam, tiagabine, and zonisamide and exhibit large interindividual variability in clearance. *CYP3A4* have more than 30 SNPs identified, and *CYP3A5* have interethnic differences.

mEH (microsomal epoxide hydrolase) metabolize carbamazepine. Interestingly, oxidation by one or more of the phase I oxidative enzymes such as the CYP superfamily often results in the

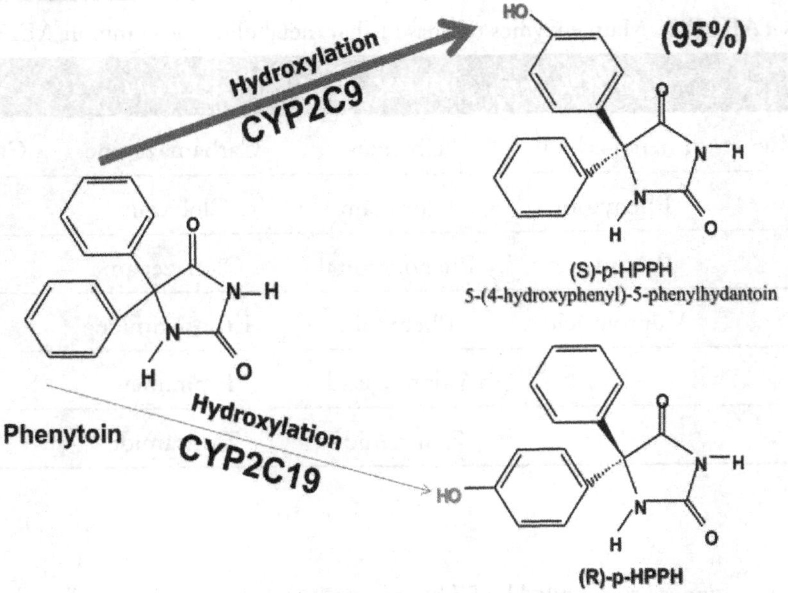

FIGURE 23: The main metabolic transformation of PHT is related with CYP2C9 activity, unless this pathway becomes saturated at higher doses of the drug.

TABLE 9: Different CYP2C9 alleles related with PHT metabolism

ALLELE	NUCLEOTIDE CHANGE	ENZYME ACTIVITY
CYP2C9*1		Active
CYP2C9*2	Arg144Cys	Intermediate
CYP2C9*3	Ile359Leu	Higher K_m and/or lower V_{max}
(Substrate dependent)		
CYP2C9*4	Ile359Thr	Intermediate; modest effect on V_{max}
CYP2C9*5	Asp360Glu	Intermediate; modest effect on V_{max}
CYP2C9*6	Adenine basepair deletion (nucleotide 818)	Inactive protein

formation of reactive xenobiotic epoxide, that are substrate of mEH encoded by *EPHX1* gene. Marked interindividual variations observed in mEH activity are due to genetic polymorphisms in both coding and promoter regions of the *EPHX1* gene [98, 99].

UGTs (UDP-glucuronosyltransferases) metabolize carbamazepine, felbamate, lamotrigine, oxcarbazepine, topiramate, valproic acid (VA), and zonisamide. The UGT enzyme activity is key in preventing the accumulation of potentially toxic lipophilic compounds and initiate their elimination through more the hydrophilic biliary and renal systems, accomplished by the addition of a hydrophilic sugar moiety (glucuronide) from UDP. Several polymorphisms have been described in UGT1 family (*UGT1A1, 1A3, 1A4, 1A6, 1A9*) and in the UGT2 family (*UGT2B4, 2B7,2B15*) [100].

The liver is the most important organ where CYP-mediated drugs metabolism occurs, making the "first modification" of orally administered therapeutics. Several CYPs isoforms exist as allelic or genetic variants, and these polymorphisms have direct influence the plasma concentration of some AEDs. In this regard, many AEDs are metabolized by *CYP2C9* and *CYP2C19*, including more commonly used AEDs such as PHT, phenobarbital, and VA, and interindividual variability was observed in patients with their CYP genotype. Furthermore, PHT metabolism is directly dependent of allelic composition of the CYP isoforms *CYP2C19* or *CYP2C9*, and mutations or gene variants of these enzymes can identify poor metabolizers from hypermetabolizers. Because CYPs are also expressed in the kidney, renal excretion of AEDs or their metabolites is an important additional elimination route that will also impact the plasmatic levels of the drugs. Additionally, functional expression of CYPs has been detected in various CNS cell types including BBB, with a differential rate of expression in distinct CNS cell populations [101].

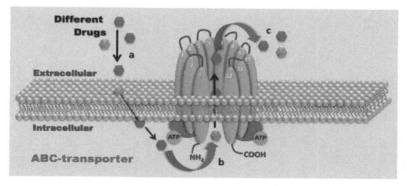

FIGURE 24: Blood circulating drugs will be transported into the hepatocytes by SLC system, metabolized by phase 1 of CYP system, and conjugated by phase 2 to be exported outside the hepatocyte into the bile. Original drugs and their metabolites can also be exported by the pumping system.

More recently, nearly 20 different isoforms of CYP were detected in brain parenchyma [102]. A differential pattern expression of these CYP isoforms in endothelial cells from epileptic tissue compared with microvascular cerebral endothelial cells was described. However, RNA expression in endothelial cells from drug-resistant epileptic patients was not different compared to brain specimens rejected from aneurysm domes without seizures [103]. These results reinforce the concept that overexpression of CYP isoforms could explain changes in PK, which require higher doses of AEDs. However, perhaps, they are not enough to explain the pharmacoresistance observed in RE patients (Figure 24).

Drug Transport System

Transporters that are important in pharmacokinetics generally are located in intestinal, renal, and hepatic epithelia. They function in the selective absorption and elimination of endogenous substances and xenobiotics, including drugs. Transporters work in concert with drug-metabolizing enzymes to eliminate drugs and their metabolites. In addition, transporters in various cell types mediate tissue-specific drug distribution (drug targeting); conversely, transporters also may serve as protective barriers to particular organs and cell types. For example, P-gp in the BBB protects the CNS from a variety of structurally diverse compounds through its efflux mechanisms. Many of the transporters that are relevant to drug response control the tissue distribution as well as the absorption and elimination of drugs.

Drug transport proteins can be grouped into two major classes, the solute carriers (SLC), divided into 48 subfamilies, and ATP-binding cassette (ABC) transporters, with approximately 19 of these gene families identified. These transporters include the organic anion transporting polypeptide (SLCO), the oligopeptide transporter (SLC15), the organic anions/cations/zwitterions transporter (SLC22), and the organic cation transporter (SLC47) families. Seven subfamilies of ABC transporter genes have been identified, encoding for 49 different proteins [104, 105].

Recently, two human orthologues of the multidrug and toxin extrusion (MATE) family of bacteria have been identified as H^+/organic cation antiport systems, and assigned as members of the SLC47 subfamily. MATE1 is primarily expressed in the kidney but also exists in the liver, adrenal gland, testis, and skeletal muscle, whereas MATE2-K has been found exclusively in the kidney [106].

SLC and ABC transporters share a wide distribution in the body and are involved in the transport of a broad range of substrates. There is growing evidence to suggest that it is the rule rather than the exception that a given drug will interact with a set of membrane transporters at some point in its deposition in the body. Several SLC and ABC transporters are currently considered to exert the greatest impact on overall drug disposition, pharmacokinetic variability, and drug–drug interactions. Intestine, liver, and kidney are the principal organs that determine the absorption, distribution, metabolism, and elimination of drugs, but the BBB plays the main role in the drugs' access to the brain.

All these organs have very similar mechanisms of drug transport. Depending on the direction in which carrier proteins translocate the substrate across the cell membrane, they can be categorized as influx or efflux transporters. ABC transporters are by definition efflux transporters because they use energy derived from ATP hydrolysis to mediate the primary active export of drugs from the intracellular to the extracellular milieu, often against a steep diffusion gradient. It is important to understand that the interplay between transporters located on apical and basolateral membranes in epithelial cells is critical in determining the extent and direction of drug movement in organs such as the intestine, liver, kidney, and BBB. The SLC subfamilies SLC15, SLC22, and SLCO are considered to have a major role in drug uptake into intestine, liver, and kidney, whereas SLC47 members mediate drug efflux into bile and urine. Many of the SLC family members facilitate the cellular uptake or influx of substrates, either by facilitated diffusion down the electrochemical gradient (acting as a channel or uniporter) or by secondary active transport against a diffusion gradient coupled to the symport or antiport of inorganic or small organic ions to provide the driving force. Certain SLC transporters exhibit efflux properties or are bidirectional, depending on the concentration gradients of substrate and coupled ion across the membrane. Few mutations in SLC transporters were related with epileptic syndrome (see above), but there were not related with refractory phenotypes.

ABC Transporters

ABC transporters are present in cellular and intracellular membranes and use energy to drive the transport of various molecules across the membranes of the cells, endoplasmic reticulum, peroxisome, as well as mitochondria [107]. There are at least 49 ABC transporter genes, which are divided into seven different families (A–G) based on sequence similarity. All families have homology of their ATP-binding regions and contain two ATP-binding regions and two transmembrane domains, except ABCG2 having one. ABC transporters are responsible for removing (efflux) of substances from cells and tissues and often transport substances against a concentration gradient by using the

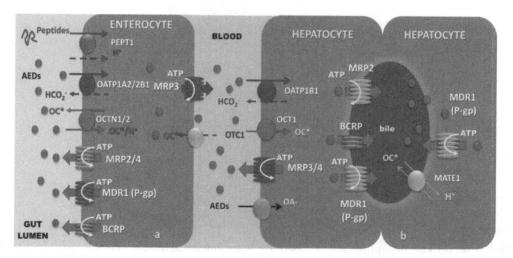

FIGURE 25: Typical structure of the ABC transporter (P-gp). (a) Different types of drugs will enter the cell through their interactions with membrane phospholipids. (b) The drugs have a high affinity to the hydrophobic drug-binding site from the inner leaflet of the membrane or the cytosol. This binding to the protein produces a conformational change in the ABC, which increases their affinity for ATP. Hydrolysis of one ATP produces a conformational change in the TMDs, which involves the movement of the helices that form the drug-binding pocket. (c) These changes decrease the drug-binding affinity and induce the delivery (efflux) of drugs outside of the cell.

hydrolysis of ATP to drive the transport. Only three of these seven gene families are particularly important for drug transport and multiple drug resistance (MDR) in tumor cells: (a) the *ABCB1* gene, encoding *MDR1* (also known as P-gp; Figure 25), (b) *ABCG2* (breast cancer resistance protein, or BCRP), and (c) the *ABCC* family or multidrug resistance proteins (MRPs).

The localization of the proteins depends on the cell type, such as hepatocyte, enterocyte, renal proximal tubule, and BBB (Figures 24, 26, and 27). The majority of ABC transporters move compounds from the cytoplasm to the outside of a cell, although some move compounds into an intercellular compartment such as the endoplasmic reticulum, mitochondria, or peroxisome.

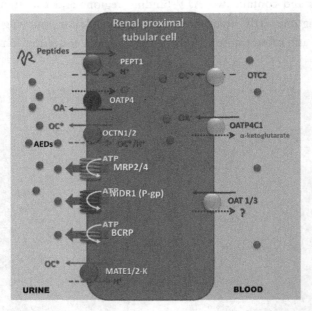

FIGURE 26: (a) The absorption of drugs from the gastrointestinal tract is a critical factor in determining oral bioavailability. Enterocytes of the small intestine are equipped with an array of influx transporters at the luminal membrane for the absorption of food components and drugs; however, as a first barrier against xenobiotics, the intestine also has a high expression of ABC transporters in the brush border membrane that can effectively pump drugs back into the intestinal lumen, thereby limiting the extent of substrate drug absorption. (b) The polarized location of these transporters in hepatocytes, facing the canalicula, ensure the excretion of xenobiotics, drug metabolites, and excess drugs.

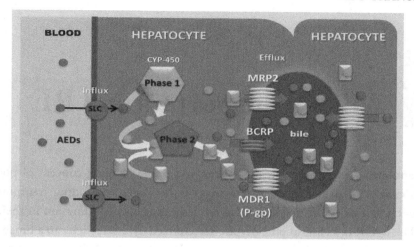

FIGURE 27: Similar polarized locations of transporters facing the external compartment of the body can be also observed in the renal tubular cells, producing the efflux of xenobiotics, drug metabolites, and drugs in excess. Altogether, these powerful metabolic and transport mechanisms will affect the final amount of drugs that impregnate the tissue and reach a steady state of concentrations in plasma, in accordance with the frequency of drug administration. Normally, 70% of an orally administered dose will be metabolized and excreted this way.

ABC Transporters and RE

The drug transporters' involvement in RE phenotype is an emerging concept of pharmacoresistance that is explained by an increased functional expression of multidrug transporter proteins, able to prevent access of AEDs to the brain, and decreasing its concentration in their sites of action [108–111]. MDR is a clinical phenotype characterized by insensitivity to a broad spectrum of drugs that presumably act on different mechanisms. Because most AEDs are administered orally, variations on genes related with drug absorption, transport, and metabolism may modify the drug plasmatic levels, body distribution, and their access to the CNS. Enterocytes and hepatocytes express the major AED-metabolizing enzymes (CYP family), and multidrug transporters such as P-gp, multidrug-resistant associated proteins (MRPs) and breast cancer resistant protein (BCRP). Their overexpression in these and other peripheral organs may play a crucial role by limiting drug absorption as well as regulating their metabolization and excretion ratio, resulting in persistent low plasmatic levels of AEDs [112, 113].

Because of their expression in transporting epithelia, including the intestine, liver, and kidney, the ABC transporters play an important role in the absorption, distribution, and removal of drugs. The functional configuration of an ABC transporter is made of two transmembrane domains, each consisting of six transmembrane helices and two cytoplasmic ATP-binding domains.

Altogether, metabolic and transport systems produce a constant elimination of the drugs, and in sum, they are the most important factor regulating the balance between dose and response, with a direct impact in the plasmatic levels of the drugs. The locations of these mechanisms in different organs will induce a unidirectional route of the drugs from the inner to the external side of the body.

MDR1 (*ABCB1*) Gene Variants

The first systematic screen of the *MDR1* gene for the presence of SNPs was published by Hoffmeyer et al. [114, 115] in 2000–2001. Of the ABC transporters, phenotypic consequences of variants have so far been most extensively described for *MDR1* and have primarily focused on the 3435C>T and 2677G>T/A SNPs (Figure 28).

However, the possible effects of individual polymorphisms on the pharmacokinetics of substrate drugs remain highly controversial. Contradictory findings may indicate that genetic variation of *MDR1* 3435C>T is not the causal modulator of any of the observed functional differences. Therefore, it is very likely that functional differences arise from SNPs in linkage disequilibrium with other (unidentified) functional polymorphism(s), including the *MDR1* 2677G>T/A polymorphism, suggesting that functional effects of genetic variants in the *MDR1* gene should be considered as haplotypes rather than independent SNPs. To date, over 100 polymorphisms that occur at a frequency of greater than 5% have been identified in Caucasians [116, 117]. Interestingly, persistent low levels of PHT was first described in a pediatric RE patient with high P-gp expression in the brain and later observed in another pediatric case, with RE and persistent low levels of VA [118, 119]; however, in both cases, polymorphisms of *MDR1* gene were not investigated.

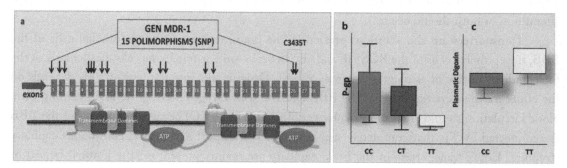

FIGURE 28: (a) Polymorphism detected on exon 26 of *MDR1* gene. Haplotypes CC are related with the (b) high expression of P-gp in the intestinal cells and (c) lower concentration of digoxin in plasma, a specific P-gp substrate.

The BBB's Role in Pharmacoresistance in Epilepsy

The CNS effects of many therapeutic drugs are blunted because of restricted entry into the brain. The basis for this poor permeability is the brain capillary endothelium, which comprises the BBB. This tissue exhibits very low paracellular (tight junction, or TJ) permeability and expresses a potent system of drug export pumps [101].

The BBB is a highly specialized structural and biochemical barrier that regulates the entry of blood-borne molecules and cells into the brain and preserves ionic homeostasis within the brain microenvironment. The structural properties of the BBB are primarily determined by the endothelial junctional complexes, consisting of TJ and adherens junctions (AdJ). The TJ complexes seal the interendothelial cleft and regulate BBB paracellular permeability, whereas the AdJ is important for initiating and maintaining endothelial cell–cell contact. This structure of the BBB plays a critical role in the control of the type and amount of therapeutics that will have access to the CNS. It is important to note that current treatment of epilepsy is based on the AED's penetration of the brain from blood. Although the BBB protects the brain from a wide spectrum of drugs, it is clear that the BBB is not impairment for the good therapeutic response in near 70% of epileptic patients. Then, a modification of BBB properties increasing the drug selectivity, intracerebal (parenchyma) alterations, or both should occur.

Transporters are abundantly expressed in the brain, mainly in the endothelial cells of the BBB, the epithelial cells of the BCSFB, and also in brain some parenchyma cells, particularly in the food-ending process of astrocytes-touching vessels. Drug uptake into the brain highly depends on the efflux transporters expressed at the BBB and BCSFB. Drugs entering the brain coming from blood circulation should cross the vascular endothelial cells (VEC) because the VECs of the BBB are connected by TJs that completely seal paracellular transport and form a continuous capillary structure. Efflux transporters in the brain capillary are mainly present on the luminal (apical or blood) side of the endothelial cells. Several models have proposed different locations of some transporters in VEC of BBB but do not give a definitive structure that can completely explain their roles in the brain protection as well as the restrictions of drugs' access to the brain parenchyma. However,

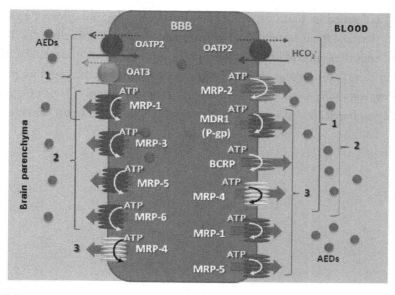

FIGURE 29: Schematic representation of a VEC in BBB and their polarized location of the drug-transporters. Note that the same transporters are described in different positions according to each author. Numbers indicate the references: (1) Bauer et al., *Exp Biol Med* 2005;230:118–127; (2) Löcher and Potschka, *NeuroRx* 2005;2,1:86–98; (3) Bollók et al., *Transworld Res Network* 2008;37/661(2):1–23.

all these models indicate that P-gp, BCRP, and MRP2 are unequivocally expressed at the blood side (Figure 29).

Consequently, these same transporters have been implicated with AED resistance. The first evidence for involvement of efflux transporters in epilepsy goes back to studies by Tishler et al. in 1995. These researchers observed increased P-gp mRNA in the brain, and protein expression in the capillary endothelium of patients with drug-resistant epilepsy [120]. The findings by Tishler et al. were confirmed by other groups [118, 121, 122], and it was suggested that this phenomenon could prevent AEDs from entering the brain and cause AED resistance. One important factor underlying AED resistance is, at least in part, seizure-induced overexpression of drug efflux transporters, not only at the BBB but also at the neurons, and these concepts will be discussed below. The brain is protected from circulating metabolites, neuroactive substances, drugs, toxins, and blood-borne pathogens by two major barriers: the BBB and the BCSFB. The BBB, formed by brain capillary endothelial cells, is characterized by highly developed TJs that restrict porous and paracellular pathways of solute diffusion from the blood into the CNS.

The BBB is formed by brain capillary endothelial cells that are joined together by TJs, the cell–cell contacts that seal the intercellular space between adjacent endothelial cells, thereby creating a nonfenestrated endothelium and limiting hydrophilic molecules from paracellular diffusion [123]. The multiprotein complexes present in the TJs guarantee a tight barrier, and thus, protection of the CNS [124]. However, under pathological conditions such as epilepsy, TJs can be dysfunctional or disrupted, leading to barrier leakage, impaired neuronal function, and brain damage [125].

Induction of ABC Transporters: Is the Acquired Refractoriness an Inducible Process?

To answer this question, we should start considering whether seizures themselves can induce the expression of some transporters.

The first evidence showing the upregulation of *mdr1* gene after seizures induced experimentally were reported by the increased P-gp expression in reactive astrocytes observed in rat brain after intracerebroventricular administration of kainite [126], in brain blood vessel endothelium, and unidentified brain cells of kainate-induced epilepsy [127] and progressively in neurons after repetitive seizures induced by 3-mercaptopropionic acid [128].

Later, in a very original and elegant experimental study [129], the expression of both *mdr1a* and *mdr1b* gene variants was investigated using GEPRs to audiogenic stimulation, avoiding any drug induction of the *mdr1* gene. The authors demonstrated that mdr1a mRNA increased and reached a maximal level at 24 h and remained increased up to 7 days after seizure induction.

Additionally, it was demonstrated that after progressive P-gp brain overexpression by repetitive induced seizures, rats were refractory to AEDs treatments, showing altered AEDs kinetics in the hippocampus and resulting in status epilepticus-associated death. However, the groups treated additionally with administration of nimodipine (2 mg/kg, ip) plus PHT reversed the refractory phenotype, recovering normal PK of PHT in hippocampus, and remained alive [129–132].

Bauer et al. [133] demonstrate that glutamate can cause localized upregulation of P-gp via cyclooxygenase 2 (COX-2), and that this phenomenon can be prevented with COX-2 inhibitors. These results indicate that P-gp upregulation in the epileptic rat brain capillaries could be prevented by the COX-2 inhibitor celecoxib. Note that Bauer at al. [133] also demonstrated that intracerebral microinjections of glutamate at nanomolar levels were sufficient to locally increase P-gp expression without seizure activity. These particular data suggest that molecular factors inducing P-gp overexpression in the BBB, even in the absence of seizures, could precondition the refractory phenotype for some epileptic syndromes. Furthermore, these findings suggest that upregulation of P-gp may be a local, but not global, effect and thus may not be able to be detected in studies that measure whole-brain uptake of drugs, producing a localized low levels of AEDs in focal epileptogenic areas, with overdoses of AEDs on normal con-convulsive brain regions. Additionally, we could speculate that the inducible property of P-gp suggests that P-gp overexpression in the brain is secondary to a wide spectrum of inductor factors, which could be the consequence of a progressive process where seizures with refractory phenotype are the result. In this context, current literature suggests that P-gp could be a new therapeutic target of drug-RE in clinical practice [134–136].

Genes and
Pharmacodynamic Modifications

Pharmacoresistance in epilepsy could be caused by the modification of one or more drug target molecules.

After the AEDs permeation into the CNS parenchyma, drugs have to bind to one or more targets to exert their desired effects. Most of AEDs predominantly target voltage-gated cation channels (the α subunits of voltage-gated Na^+ channels and also T-type voltage-gated Ca^{2+} channels) or influence GABA-mediated inhibition.

Voltage-gated sodium channels, the structure and genetics of which were described above, have nine different a subunits and four b subunits identified, and the a subunits are main functional Na^+ channel components (see Figure 6). Of the nine mammalian genes encoding α subunits of active channels, four (SCN1A, SCN2A, SCN3A, and SCN8A) are expressed in the CNS.

It has been suggested that altered composition, function, and/or expression of different subunits of Na^+ channels are related with pharmacoresistant epilepsy. Interestingly, Tate et al. [137] assessed whether variation in SCN1A in patients with epilepsy was associated with the clinical use of carbamazepine and PHT. They found that patients with AA genotype of one SNP, namely rs3812718 or SCN1A IVS5-91, more frequently received higher maximum doses of both carbamazepine and PHT. According to these data, as the need for higher AED dosages implicates a worsened reaction on AEDs, patients carrying the SCN1A IVS5–91 AA genotype have a higher risk for drug-resistant epilepsy. All these alterations could result in the expression of a phenotype of AED-insensitive SCN1A α subunit. However, in the studies by Gombert-Handoko et al. [138], no association was found between SCN1A IVS5-91 G>A polymorphism and drug-resistant epilepsy.

One interesting pharmacodynamic change is observed after an aberrant bursting in CA1 hippocampal neurons from epileptic animals mediated by an increased expression of T-type Ca^{2+} channels [139] or in thalamic neurons implicated in the generation of spike-wave discharges in absence of epilepsy [140]. This type of mechanism could be also applied to other voltage-gated ion channels such as K^+ channels [108]. In humans, these types of modifications that reduce efficacy of a given AED at the "target" level were described on voltage-gated sodium channels by

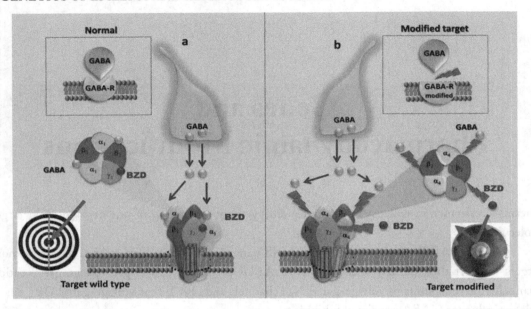

FIGURE 30: (a) Normal recognition of GABA and BZD on their binding sites of GABA-R. (b) Changes on the expression of the isoforms of a subunits induce the loss of GABA and BDZ affinity on their binding sites, thus reducing the efficacy of their inhibitory signaling.

downregulation of their accessory β subunits, altered α subunit expression, or induction of neonatal sodium channel II and III α-isoform mRNAs [141]. Although the *SCN1A* a subunit gene has been the focus of most research so far, the expression of *SCN2A*, *SCN3A*, *SCN3B*, and *SCN8A* was also observed altered in patients with RE [55].

Similar changes were observed in the $GABA_A$ receptors: through a decrease in a1 subunits and increase in a4 subunits, the GABA and BZDs' affinity for their receptor is reduced (Figure 30). These mechanisms resulted in modifications of specific "targets" associated with seizure activity, producing changes at transcription level or alternative splicing on mRNA of ion channel subunits, as well as altered posttranslational modification of the protein and/or phosphorylation by protein kinases.

One intriguing question is that while carbamazepine, PHT, valproate, and lamotrigine bind to the same target (Na^+ channels) [142], the reduced pharmacosensitivity to these drugs following pilocarpine-induced status epilepticus in rats depends on the AED selected [143].

Irrespective of the fact that the upregulation and downregulation of several different subunits of ion channels can be related to modification of AED efficacy, mutations on these subunits were related not only to the severity and type of epileptic syndrome but also with the pharmacoresistance in epilepsy. As an example, mutations of the b1 subunit of Na$^+$ channels are the cause of the epilepsy syndrome, generalized epilepsy with febrile seizures plus [144]. Interestingly, the mutant b1 subunit of this channel is associated with a dramatic and selective loss of use-dependent block by PHT [144] and carbamazepine [143–145]. Collectively, all of the pharmacodynamic modifications resulting in the lost of sensitivity or increased refractoriness has been termed "the target hypothesis of pharmacoresistance [108]."

Inducing the Expression of P-gp in Neurons: Is it to Induce Epileptogenesis?

Finally, a very intriguing concept that should be addressed regarding the pharmacoresistant phenotype associated with pharmacodynamic changes is the potential role of P-gp and perhaps others ABC transporters in the modification of ion balance and neuronal membrane polarization. An alternative mechanism to the classic pumping function of P-gp was observed in cells expressing the *MDR1* gene. These cells exhibit significantly lower membrane potential ($\Delta\Psi0 = -10$ to -20 mV) compared with the physiological potential ($\Delta\psi0$ of -60 mV), leading to reduced (approximately 30%) binding of the drug [146]. In neurons, these P-gp-dependent potential membrane alterations ($\Delta\psi0$) can contribute not only to the development of the refractory phenotype but also to the intrinsic mechanisms of the epileptogenicity. In agreement with these concepts, a preliminary collaborative study showed the first evidence that repetitive seizures induce high neuronal P-gp overexpression associated with refractoriness and a concomitant progressive enrollment of hippocampal cells with depolarized membranes. Both refractoriness and depolarization were reversed only after simultaneous nimodipine and PHT administration. Irrespective of the well-known drug transport property, induced neuronal P-gp overexpression from many different causes could be an additional mechanism of membrane depolarization, which increases the risk for new seizures and thus plays a role in epileptogenesis (Figure 31) [147].

In accordance with this concept, it is important to regard the wide spectrum of factors that can upregulate P-gp expression, including brain hypoxia, brain tumors, inflammation, autoimmunity, vascular malformations, brain infections, brain trauma, metabolic disorders, stroke, and seizure itself.

Interestingly, all these conditions not only share the epilepsy as indicated in Figure 13 but also share the pharmacoresistant phenotype.

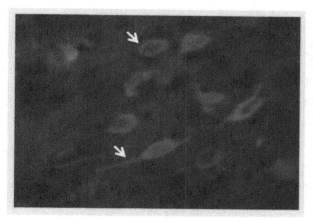

FIGURE 31: P-gp immunostaining in cortical neurons. Which could be the functional changes in neurons expressing P-gp, taking in account that in a normal brain, neurons do not express P-gp?

References

[1]   Glauser TA, Sankar R. Core elements of epilepsy diagnosis and management: expert consensus from the Leadership in Epilepsy, Advocacy, and Development (LEAD) faculty. *Curr Med Res Opin* 2008;24:3463–77.

[2]   World Health Organization Epilepsy, 2009. Fact Sheet No. 999. WHO.

[3]   Fisher RS, van Emde Boas W, Blume W, Elger C, Genton P, Lee P, Engel J Jr. Epileptic seizures and epilepsy: definitions proposed by the International League Against Epilepsy (ILAE) and the International Bureau for Epilepsy (IBE). *Epilepsia* 2005;46:470–2.

[4]   Steinlein OK, Mulley JC, Propping P, Wallace RH, Phillips HA, Sutherland GR, Scheffer IE, Berkovic SF. A missense mutation in the neuronal nicotinic acetylcholine receptor α4 subunit is associated with autosomal dominant nocturnal frontal lobe epilepsy. *Nat Genet* 1995;11:201–3.

[5]   Phillips HA, Scheffer IE, Berkovic SF, Hollway GE, Sutherland GR, Mulley JC. Localization of a gene for autosomal dominant nocturnal frontal lobe epilepsy to chromosome 20q13.2. *Nat Genet* 1995;10:117–8.

[6]   Hayman M, Scheffer IE, Chinvarun Y, Berlangieri SU, Berkovic SF. Autosomal dominant nocturnal frontal lobe epilepsy: demonstration of focal frontal onset and intrafamilial variation. *Neurology* 1997;49:969–75.

[7]   Engelborghs S, D'Hooge R, De Deyn PP. Pathophysiology of epilepsy. *Acta Neurol Belg* 2000;100:201–13.

[8]   Shorvon SD. The etiologic classification of epilepsy. *Epilepsia* 2011;52:1052–7.

[9]   Ferrie CD. Terminology and organization of seizures and epilepsies: radical changes not justified by new evidence. *Epilepsia* 2010;51:713–4.

[10]   Guerrini R. Classification concepts and terminology: is clinical description assertive and laboratory testing objective? *Epilepsia* 2010;51:718–20.

[11]   Wolf P. Much ado about nothing? *Epilepsia* 2010;51:717–8.

[12]   Rogawski MA, Johnson MR. Intrinsic severity as a determinant of antiepileptic drug refractoriness. *Epilepsy Curr* 2008;8:127–30.

[13]   Lodish H, Berk A, Zipursky S, Matsudaira P, Baltimore D, Darnell J. *Molecular Cell Biology*. 4th ed. New York: W.H. Freeman; 2000.

[14]   Meldrum BS, Rogawski MA. Molecular targets for antiepileptic drug development. *Neuro-therapeutics* 2007;4:18–61.

[15]   Rogawski MA. Revisiting AMPA receptors as an antiepileptic drug target. *Epilepsy Curr* 2011;11:56–63.

[16]   Tan NC, Mulley JC, Berkovic SF. Genetic association studies in epilepsy: the truth is out there. *Epilepsia* 2004;45:1429–42.

[17]   Heinzen EL, Depondt C, Cavalleri GL, Ruzzo EK, Walley NM, Need AC, Ge D, He M, Cirulli ET, Zhao Q, Cronin KD, Gumbs CE, Campbell CR, Hong LK, Maia JM, Shianna KV, McCormack M, Radtke RA, O'Conner GD, Mikati MA, Gallentine WB, Husain AM, Sinha SR, Chinthapalli K, Puranam RS, McNamara JO, Ottman R, Sisodiya SM, Delanty N, Goldstein DB. Exome sequencing followed by large-scale genotyping fails to identify single rare variants of large effect in idiopathic generalized epilepsy. *Am J Hum Genet* 2012;91:293–302.

[18]   Reid CA, Berkovic SF, Petrou S. Mechanisms of human inherited epilepsies. *Prog Neurobiol* 2009;87:41–57.

[19]   Singh R, Gardner RJ, Crossland KM, Scheffer IE, Berkovic SF. Chromosomal abnormalities and epilepsy: a review for clinicians and gene hunters. *Epilepsia* 2002;43:127–40.

[20]   Berkovic SF, Howell RA, Hay DA, Hopper JL. Epilepsies in twins: genetics of the major epilepsy syndromes. *Ann Neurol* 1998;43:435–45.

[21]   Ottman R, Annegers JF, Risch N, Hauser WA, Susser M. Relations of genetic and environmental factors in the etiology of epilepsy. *Ann Neurol* 1996;39:442–9.

[22]   Ottman R. Analysis of genetically complex epilepsies. *Epilepsia* 2005;46(Suppl 10):7–14.

[23]   Andermann F, Kobayashi E, Andermann E. Genetic focal epilepsies: state of the art and paths to the future. *Epilepsia* 2005;46(Suppl 10):61–7.

[24]   Hauser WA, Annegers JF, Kurland LT. Incidence of epilepsy and unprovoked seizures in Rochester, Minnesota: 1935–1984. *Epilepsia* 1993 May–Jun;34:453–68.

[25]   Hesdorffer DC, Logroscino G, Cascino G, Annegers JF, Hauser WA. Incidence of status epilepticus in Rochester, Minnesota, 1965–1984. *Neurology* 1998;50:735–41.

[26]   Berkovic SF and Scheffer IE. Genetics of the epilepsies. *Curr Opin Neurol* 1999;12:177–82.

[27]   Ottman R, Lee JH, Hauser WA, Risch N. Are generalized and localization-related epilepsies genetically distinct? *Arch Neurol* 1998;55:339–44.

[28]   Risch NJ. Searching for genetic determinants in the new millennium. *Nature* 2000;405:847–56.

[29]   Combi R, Dalpra L, Tenchini ML, Ferini-Strambi L. Autosomal dominant nocturnal frontal lobe epilepsy—a critical overview. *J Neurol* 2004;251:923–34.

[30] Scheffer IE, Bhatia KP, Lopes-Cendes I, et al. Autosomal dominant nocturnal frontal lobe epilepsy. A distinctive clinical disorder. *Brain* 1995;118:61–73.

[31] Aridon P, Marini C, Di Resta C, et al. Increased sensitivity of the neuronal nicotinic receptor alpha 2 subunit causes familial epilepsy with nocturnal wandering and ictal fear. *Am J Hum Genet* 2006;79:342–50.

[32] Wallace RH, Wang DW, Singh R, Scheffer IE, George AL Jr, Phillips HA, Saar K, Reis A, Johnson EW, Sutherland GR, Berkovic SF, Mulley JC. Febrile seizures and generalized epilepsy associated with a mutation in the Na$^+$-channel beta1 subunit gene SCN1B. *Nat Genet* 1998;19:366–70.

[33] Audenaert D, Claes L, Ceulemans B, Löfgren A, Van Broeckhoven C, De Jonghe P. A deletion in SCN1B is associated with febrile seizures and early-onset absence epilepsy. *Neurology* 2003;61:854–6.

[34] Escayg A, MacDonald BT, Meisler MH, Baulac S, Huberfeld G, An-Gourfinkel I, Brice A, LeGuern E, Moulard B, Chaigne D, Buresi C, Malafosse A Mutations of SCN1A, encoding a neuronal sodium channel, in two families with GEFS+2. *Nat Genet* 2000;24:343–5.

[35] Claes L, Del-Favero J, Ceulemans B, Lagae L, Van Broeckhoven C, De Jonghe P. De novo mutations in the sodium-channel gene SCN1A cause severe myoclonic epilepsy of infancy. *Am J Hum Genet* 2001;68:1327–32.

[36] Meisler MH, Kearney JA. Sodium channel mutations in epilepsy and other neurological disorders. *J Clin Invest* 2005;115:2010–7.

[37] Spampanato J, Kearney JA, de Haan G, McEwen DP, Escayg A, Aradi I, MacDonald BT, Levin SI, Soltesz I, Benna P, Montalenti E, Isom LL, Goldin AL, Meisler MH. A novel epilepsy mutation in the sodium channel SCN1A identifies a cytoplasmic domain for beta subunit interaction. *J Neurosci* 2004;24:10022–34.

[38] McIntosh AM, McMahon J, Dibbens LM, Iona X, Mulley JC, Scheffer IE, Berkovic SF. Effects of vaccination on onset and outcome of Dravet syndrome: a retrospective study. *Lancet Neurol* 2010;9:592–8.

[39] Emond MR, Biswas S, Jontes JD. Protocadherin-19 is essential for early steps in brain morphogenesis. *Dev Biol* 2009;334(1):72–83.

[40] Dibbens LM, Tarpey PS, Hynes K, Bayly MA, Scheffer IE, Smith R, Bomar J, Sutton E, Vandeleur L, Shoubridge C, Edkins S, Turner SJ, Stevens C, O'Meara S, Tofts C, Barthorpe S, Buck G, Cole J, Halliday K, Jones D, Lee R, Madison M, Mironenko T, Varian J, West S, Widaa S, Wray P, Teague J, Dicks E, Butler A, Menzies A, Jenkinson A, Shepherd R, Gusella JF, Afawi Z, Mazarib A, Neufeld MY, Kivity S, Lev D, Lerman-Sagie T, Korczyn AD, Derry CP, Sutherland GR, Friend K, Shaw M, Corbett M, Kim HG, Geschwind DH, Thomas P, Haan E, Ryan S, McKee S, Berkovic SF, Futreal PA, Stratton

MR, Mulley JC, Gécz J. X linked protocadherin 19 mutations cause female-limited epilepsy and cognitive impairment. *Nat Genet* 2008;40:776–81.

[41]   Depienne C, LeGuern E. PCDH19-related infantile epileptic encephalopathy: an unusual X-linked inheritance disorder. *Hum Mutat* 2012;33:627–34.

[42]   Terracciano A, Specchio N, Darra F, Sferra A, Bernardina BD, Vigevano F, Bertini E. Somatic mosaicism of PCDH19 mutation in a family with low-penetrance EFMR. *Neurogenetics* 2012;13:341–5.

[43]   Biervert C, Schroeder BC, Kubisch C, Berkovic SF, Propping P, Jentsch TJ, Steinlein OK. A potassium channel mutation in neonatal human epilepsy. *Science* 1998;279:403–6.

[44]   Charlier C, Singh NA, Ryan SG, Lewis TB, Reus BE, Leach RJ, Leppert M. A pore mutation in a novel KQT like potassium channel gene in an idiopathic epilepsy family. *Nat Genet* 1998;18:53–5.

[45]   Singh NA, Charlier C, Stauffer D, DuPont BR, Leach RJ, Melis R, Ronen GM, Bjerre I, Quattlebaum T, Murphy JV, McHarg ML, Gagnon D, Rosales TO, Peiffer A, Anderson VE, Leppert M. A novel potassium channel gene, KCNQ2, is mutated in an inherited epilepsy of newborns. *Nat Genet* 1998;18:25–9.

[46]   Dedek K, Fusco L, Teloy N, Steinlein OK. Neonatal convulsions and epileptic encephalopathy in an Italian family with a missense mutation in the fifth transmembrane region of KCNQ2. *Epilepsy Res* 2003;54:21–7.

[47]   Rivera C, Voipio J, Payne JA, Ruusuvuori E, Lahtinen H, Lamsa K, Pirvola U, Saarma M, Kaila K. The K^+/Cl^- co-transporter KCC2 renders GABA hyperpolarizing during neuronal maturation. *Nature* 1999;397:251–5.

[48]   Shieh C-C, Coglham M, Sullivan J, Gopalakrishnam M. Potassium channels: molecular defects, diseases, and therapeutic opportunities. *Pharmacol Rev* 2000;52:557–93.

[49]   Heron SE, Cox K, Grinton BE, Zuberi SM, Kivity S, Afawi Z, Straussberg R, Berkovic SF, Scheffer IE, Mulley JC. Deletions or duplications in KCNQ2 can cause benign familial neonatal seizures. *J Med Genet* 2007;44:791–6.

[50]   Steinlein OK, Conrad C, Weidner B. Benign familial neonatal convulsions: always benign? *Epilepsy Res* 2007;73:245–9.

[51]   Sills GJ. The mechanisms of action of gabapentin and pregabalin. *Curr Opin Pharmacol* 2006;6:108–13.

[52]   Borgatti R, Zucca C, Cavallini A, Ferrario M, Panzeri C, Castaldo P, Soldovieri MV, Baschirotto C, Bresolin N, Dalla Bernardina B, Taglialatela M, Bassi MT. A novel mutation in KCNQ2 associated with BFNC, drug resistant epilepsy, and mental retardation. *Neurology* 2004;63:57–65.

[53]   Khosravani H, Bladen C, Parker DB, Snutch TP, McRory JE, Zamponi GW. Effects of Cav3.2 channel mutations linked to idiopathic generalized epilepsy. *Ann Neurol* 2005;57:745–9.

[54]   Vitko I, Chen Y, Arias JM, Shen Y, Wu XR, Perez-Reyes, E. Functional characterization and neuronal modelling of the effects of childhood absence epilepsy variants of CACNA1H, a T-type calcium channel. *J Neurosci* 2005;25:4844–55.

[55]   Baulac S, Huberfeld G, Gourfinkel-An I, et al. First genetic evidence of GABA(A) receptor dysfunction in epilepsy: a mutation in the gamma2-subunit gene. *Nat Genet* 2001;28: 46–8.

[56]   Wallace RH, Marini C, Petrou S, Harkin LA, Bowser DN, Panchal RG, Williams DA, Sutherland GR, Mulley JC, Scheffer IE, Berkovic SF. Mutant GABAA receptor γ2-subunit in childhood absence epilepsy and febrile seizures. *Nat Genet* 2001;28:49–52.

[57]   Cossette P, Liu L, Brisebois K, Dong H, Lortie A, Vanasse M, Saint-Hilaire JM, Carmant L, Verner A, Lu WY, Wang YT, Rouleau GA. Mutation of GABRA1 in an autosomal dominant form of juvenile myoclonic epilepsy. *Nat Genet* 2002;31:184–9.

[58]   Maljevic S, Krampfl K, Cobilanschi J, Tilgen N, Beyer S, Weber YG, Schlesinger F, Ursu D, Melzer W, Cossette P, Bufler J, Lerche H, Heils A. A mutation in the GABAA receptor 1-subunit is associated with absence epilepsy. *Ann Neurol* 2006;59:983–7.

[59]   Dibbens LM, Feng HJ, Richards MC, Harkin LA, Hodgson BL, Scott D, Jenkins M, Petrou S, Sutherland GR, Scheffer IE, Berkovic SF, Macdonald RL, Mulley JC. GABRD encoding a protein for extra- or peri-synaptic GABAA receptors is a susceptibility locus for generalized epilepsies. *Hum Mol Genet* 2004;13(13):1315–19

[60]   Macdonald RL, Kang JQ, Gallagher JM. Mutations in GABAA receptor subunits associated with genetic epilepsies. *J Physiol* 2010;88(11):1861–9 1861.

[61]   Lachance-Touchette P, Brown P, Meloche C, Kinirons P, Lapointe L, Lacasse H, Lortie A, Carmant L, Bedford F, Bowie D, Cossette P. Novel α1 and χ2 GABAA receptor subunit mutations in families with idiopathic generalized epilepsy. *Eur J Neurosci* 2011;34:237–49.

[62]   Saint-Martin C, Gauvain G, Teodorescu G, Gourfinkel-An I, Fedirko E, Weber YG, Maljevic S, Ernst JP, Garcia-Olivares J, Fahlke C, Nabbout R, LeGuern E, Lerche H, Poncer JC, Depienne C. Two novel CLCN2 mutations accelerating chloride channel deactivation are associated with idiopathic generalized epilepsy. *Hum Mutat* 2009;30:397–405.

[63]   Oldham MC, Konopka G, Iwamoto K, Langfelder P, Kato T, Horvath S, Geschwind DH. Functional organization of the transcriptome in human brain. *Nat Neurosci* 2008;11:271–82.

[64]   Johnson MR, Shorvon SD. Heredity in epilepsy: neurodevelopment, comorbidity, and the neurological trait. *Epilepsy Behav* 2011;22:421–7.

[65]   Scheffer I, Berkovic S. Copy number variants—an unexpected risk factor for the idiopathic generalized epilepsies. *Brain* 2010:133;7–8.

[66]   Sharp AJ, Mefford HC, Li K, Baker C, Skinner C, Stevenson RE, et al. A recurrent 15q13.3 microdeletion syndrome associated with mental retardation and seizures. *Nat Genet* 2008;40:322–8.

[67]   Mulley JC, Scheffer IE, Harkin LA, Berkovic SF, Dibbens LM. Susceptibility genes for complex epilepsy. *Human Mol Gen* 2005;14:R243–9.

[68]   N'Gouemo P, Yasuda RP, Faingold CL. Seizure susceptibility is associated with altered protein expression of voltage-gated calcium channel subunits in inferior colliculus of the genetically epilepsy-prone rat. *Brain Res* 2010;1308:153–7.

[69]   Brooks-Kayal A. Molecular mechanisms of cognitive and behavioral comorbidities of epilepsy in children. *Epilepsia* 2011;52(Suppl 1):13–20.

[70]   Dolen G, Bear MF. Role for metabotropic glutamate receptor 5 (mGluR5) in the pathogenesis of fragile X syndrome. *J Physiol* 2008;586(6):1503–8.

[71]   Pin J-P, Acher F. The metabotropic glutamate receptors: structure, activation mechanism and pharmacology. *Curr Drug Targets CNS Disord* 2002;1:297–317.

[72]   Conn PJ. Physiological roles and therapeutic potential of metabotropic glutamate receptors. *Ann N Y Acad Sci* 2003;1003:12–21.

[73]   Schoepp DD. Unveiling the functions of presynaptic metabotropic glutamate receptors in the central nervous system. *J Pharmacol Exp Ther* 2001;299:12–20.

[74]   Sansig G, Bushell TJ, Clarke VR, Rozov A, Burnashev N, Portet C, Gasparini F, Schmutz M, Klebs K, Shigemoto R, Flor PJ, Kuhn R, Knoepfel T, Schroeder M, Hampson DR, Collett VJ, Zhang C, Duvoisin RM, Collingridge GL, van Der Putten H. Increased seizure susceptibility in mice lacking metabotropic glutamate receptor 7. *J Neurosci* 2001;21:8734–45.

[75]   Akiyama K, Daigen A, Yamada N, Itoh T, Kohira I, Ujike H, Otsuki S. Long-lasting enhancement of metabotropic excitatory amino acid receptor-mediated polyphosphoinositide hydrolysis in the amygdala/pyriform cortical kindled rats. *Brain Res* 1992;569:71–7.

[76]   Keele NB, Zinebi F, Neugebauer V, Shinnick-Gallagher P. Epileptogenesis upregulates metabotropic glutamate receptor activation of sodium-calcium exchange current in the amygdala. *J Neurophysiol* 2000;83:2458–62.

[77]   Pacheco Otalora LF, Couoh J, Shigamoto R, Zarei MM, Garrido Sanabria ER. Abnormal mGluR2/3 expression in the perforant path termination zones and mossy fibers of chronically epileptic rats. *Brain Res* 2006;1098:170–85.

[78]   Klapstein GJ, Meldrum BS, Mody I. Decreased sensitivity to group III mGluR agonists in the lateral perforant path following kindling. *Neuropharmacology* 1999;38:927–33.

[79]   Princivalle AP, Richards DA, Duncan JS, Spreafico R, Bowery NG. Modification of GABAB1 and GABAB2 receptor subunits in the somatosensory cerebral cortex and thalamus of rats with absence seizures (GAERS). *Epilepsy Res* 2003;55:39–51.

[80]   Tzoulis Ch, Bindoff LA. The syndrome of mitochondrial spinocerebellar ataxia and epilepsy caused by POLG mutations. *ACNR* 2009;9(3):13–6.

[81]   Ottman R, Risch N, Allen Hauser W, Pedley TA, Lee JH, Barker-Cummings Ch, Lustenberger A, Nagle KJ, Lee KS, Scheuer ML, Neystat M, Susser M, Wilhelmsen KC. Localization of a gene for partial epilepsy to chromosome 10q. *Nat Genet* 1995;10:56–60.

[82]   Brodtkorb E, Gu W, Nakken KO, Fischer C, Steinlein OK. Familial temporal lobe epilepsy with aphasic seizures and linkage to chromosome 10q22–q24. *Epilepsia* 2002;43:228–35.

[83]   Kalachikov S, Evgrafov O, Ross B, Winawer M, Barker-Cummings C, Martinelli Boneschi F, Choi C, Morozov P, Das K, Teplitskaya E, Yu A, Cayanis E, Penchaszadeh G, Kottmann AH, Pedley TA, Hauser WA, Ottman R, Gilliam TC. Mutations in LGI1 cause autosomal dominant partial epilepsy with auditory features. *Nat Genet* 2002;30:335–41.

[84]   Gu W, Brodtkorb E, Steinlein OK. LGI1 is mutated in familial temporal lobe epilepsy characterized by aphasic seizures. *Ann Neurol* 2002;52:364–7.

[85]   Gu W, Wevers A, Schröder H, Grzeschik KH, Derst C, Brodtkorb E, de Vos R, Steinlein OK. The LGI1 gene involved in lateral temporal lobe epilepsy belongs to a new subfamily of leucine-rich repeat proteins. *FEBS Lett* 2002;519:71–6.

[86]   Nakayama J, Fu YH, Clark AM, Nakahara S, Hamano K, Iwasaki N, Matsui A, Arinami T, Ptácek LJ. A nonsense mutation of the MASS1 gene in a family with febrile and afebrile seizures. *Ann Neurol* 2002;52:654–7.

[87]   Youssef-Turki IH, Kraoua I, Smirani S, Mariem K, BenRhouma H, Rouissi A, Gouider-Khouja N. Epilepsy Aspects and EEG patterns in neuro-metabolic diseases. *J Behav Brain Sci* 2011;1:69–74.

[88]   Wolf NI, Bast T, Surtees R. Epilepsy in inborn errors of metabolism. *Epileptic Disord* 2005;7:67–81.

[89]   Staretz-Chacham O, Lang TC, LaMarca ME, Krasnewich D, Sidransky E. Lysosomal storage disorders in the newborn. *Pediatrics* 2009;123:1191–207.

[90]   Xinhan L, Matsushita M, Numaza M, Taguchi A, Mitsui K, and Kanazawa H. Na+/H+ exchanger isoform 6 (NHE6/SLC9A6) is involved in clathrin-dependent endocytosis of transferrin. *Am J Physiol Cell Physiol* December 1, 2011;301(6):C1431–44.

[91]   Cavanagh NP, Bicknell J, Howard F. Cystinuria with mental retardation and paroxysmal dyskinesia in 2 brothers. *Arch Dis Child* 1974;49:662–4.

[92]   Elger C. Pharmacoresistance: modern concept and basic data derived from human brain tissue. *Epilepsia* 2003;44(Suppl 5):9–15.

[93]   Kwan P, Brodie MJ. Early identification of refractory epilepsy. *N Engl J Med* 2000;342: 314–9.

[94]   Sillanpää M, Schmidt D. Natural study of treated childhood-onset epilepsy: prospective long-term population-based study. *Brain* 2006;129(Pt 3):617–24.

[95]   Sisodiya SM. Genetics of drug resistance. *Epilepsia* 2005;46(Suppl 10):33–8.

[96]   Sisodiya SM. Genetics and Epilepsy. *ACNR* 2006;5(6):10–1.

[97]   Mosher CM, Tai G, Rettie AE. CYP2C9 amino acid residues influencing phenytoin turnover and metabolite regio- and stereochemistry. *J Pharmacol Exp Ther* 2009:329:938–44.

[98]   Hines RN, Koukouritaki SB, Poch MT, Stephens MC. Regulatory polymorphisms and their contribution to interindividual differences in the expression of enzymes influencing drug and toxicant disposition. *Drug Metab Rev* 2008;40:263–301.

[99]   Nakajima Y, Saito Y, Shiseki K, Fukushima–Uesaka H, Hasegawa R, Ozawa S, Sugai K, Katoh M, Saitoh O, Ohnuma T, Kawai M, Ohtsuki T, Suzuki C, Minami N, Kimura H, Goto Y, Kamatani N, Kaniwa N, Sawada J. Haplotype structures of EPHX1 and their effects on the metabolism of carbamazepine-10,11-epoxide in Japanese epileptic patients. *Eur J Clin Pharmacol* 2005;61:25–34.

[100]   Saruwatari J, Ishitsu T, Nakagawa K. Update on the genetic polymorphisms of drug-metabolizing enzymes in antiepileptic drug therapy. *Pharmaceuticals* 2010;3:2709–32.

[101]   Bauer B, Schlichtiger J, Pekcec A, Hartz A. Chapter 2: The blood–brain barrier in epilepsy. In: *Clinical and Genetic Aspects of Epilepsy*. Edited by Afawi Z. Intech. Pag 23–54. ISBN: 978-953-307-700-0- Source: InTech (2011).

[102]   Dutheil F, Jacob A, Dauchy S, Beaune P, Scherrmann JM, Declèves X, Loriot MA. ABC transporters and cytochromes P450 in the human central nervous system: influence on brain pharmacokinetics and contribution to neurodegenerative disorders. *Expert Opin Drug Metab Toxicol* 2010;6(10):1161–74.

[103]   Ghosh C, Gonzalez-Martinez J, Hossain M, Cucullo L, Fazio V, Janigro D, Marchi N. Pattern of P450 expression at the human blood–brain barrier: roles of epileptic condition and laminar flow. *Epilepsia* 2010;8:1408–17.

[104]   Dean M and Allikmets R. Complete characterization of the human ABC gene family. *J Bioenerg Biomembr* 2001;33:475–9.

[105]   Sheps JA and Ling V. Preface: the concept and consequences of multidrug resistance. *Pflugers Arch* 2007;453:545–53.

[106]   Frans G.M. Russel. Transporters: importance in drug absorption, distribution, and removal. (Chapter 2): In: *Enzyme- and Transporter-Based Drug–Drug Interactions. Progress and Future Challanges*. Edited by Pang KS, Rodríguez AD, Peter RM. American Association of Pharmaceutical Scientists, 2010:27–49.

[107]   Dean M, Hamon Y, Chimini G. The human ATP-binding cassette (ABC) transporter superfamily. *J Lipid Res* 2001;42:1007–17.

[108]   Remy S, Beck H. Molecular and cellular mechanisms of pharmacoresistance in epilepsy. *Brain* 2006;129(Pt 1):18–35.

[109]   Lazarowsk A, Czornyj L, Lubieniecki F, Vazquez S, D'Giano C, Sevlever G, Taratuto AL, Brusco A, Girardi E. Multidrug-resistance (MDR) proteins develops refractory epilepsy phenotype: clinical and experimental evidences. *Curr Drug Ther* 2006;1(3):291–309.

[110]   Löscher W, Sills GJ. Drug resistance in epilepsy: why is a simple explanation not enough? *Epilepsia* 2007;48:2370–2.

[111]   Potschka H. Transporter hypothesis of drug-resistant epilepsy: challenges for pharmaco-genetic approaches. *Pharmacogenomics* 2010;11:1427–38.

[112]   Lazarowski A, Lubieniecki F, Camarero S, Pomata H, Bartuluchi M, Sevlever G, Taratuto AL. Multidrug resistance proteins in tuberous sclerosis and refractory epilepsy. *Pediatr Neurol* 2004;30:102–6.

[113]   Lazarowski L and Czornyj A. Potential role of multidrug resistant proteins in refractory epilepsy and antiepileptic drugs interactions. *Drug Metab Drug Interact* 2011;26:21–6.

[114]   Hoffmeyer S, Burk O, von Richter O, Arnold HP, Brockmöller J, Johne A, Cascorbi I, Gerloff T, Roots I, Eichelbaum M, Brinkmann U. Functional polymorphisms of the human multidrugresistance gene: multiple sequence variation and correlation of one allele with P-glycoprotein expression and activity in vivo. *Proc Natl Acad Sci USA* 2000;97:3473–8.

[115]   Hoffmeyer S, Brinkmann U, Cascorbi I. Frequency of single nucleotide polymorphisms in the P-glycoprotein drug transporter MDR1 gene in white subjects. *Pharmacogenomics* 2001;2:51–64.

[116]   Saito S, Iida A, Sekine A, Miura Y, Ogawa C, Kawauchi S, Higuchi S, Nakamura Y. Three hundred twenty-six genetic variations in genes encoding nine members of ATP-, subfamily B (ABCB/MDR/TAP), in the Japanese population. *J Hum Genet* 2002;47(1):38–50.

[117]   Schwab M, Eichelbaum M, Fromm MF. Genetic polymorphisms of the human MDR1 drug transporter. *Annu Rev Pharmacol Toxicol* 2003;43:285–307.

[118]   Lazarowski A, Sevlever G, Taratuto A, Massaro M, Rabinowicz A. Tuberous sclerosis associated with MDR1 gene expression and drug-resistant epilepsy. *Pediatr Neurol* 1999;21:731–4.

[119]   Lazarowski A, Massaro M, Schteinschnaider A, Intruvini S, Sevlever G, Rabinowicz A. Neuronal MDR-1 gene expression and persistent low levels of anticonvulsants in a child with refractory epilepsy. *Ther Drug Monit* 2004;26(1):44–6.

[120]   Tishler DM, Weinberg KI, Hinton DR, Barbaro N, Annett GM, Raffel C. MDR1 gene expression in brain of patients with medically intractable epilepsy. *Epilepsia* 1995;36:1–6.

[121] Dombrowski SM, Desai SY, Marroni M, Cucullo L, Goodrich K, Bingaman W, Mayberg MR, Bengez L, Janigro D. Overexpression of multiple drug resistance genes in endothelial cells from patients with refractory epilepsy. *Epilepsia* 2001;42:1501–6.

[122] Sisodiya SM, Lin WR, Harding BN, Squier MV, Thom M. Drug resistance in epilepsy: expression of drug resistance proteins in common causes of refractory epilepsy. *Brain* 2002;125(Pt 1):22–31.

[123] Nag, S. Morphology and molecular properties of cellular components of normal cerebral vessels. *Methods Mol Med* 2003;89:3–36.

[124] Kniesel U and Wolburg H. Tight junctions of the blood–brain barrier. *Cell Mol Neurobiol* 2000;20(1):57–76.

[125] Huber JD, Egleton RD, Davis TP. Molecular physiology and pathophysiology of tight junctions in the blood–brain barrier. *Trends Neurosci* 2001;24(12):719–25.

[126] Zhang L, Ong W, Lee T. Induction of P-glycoprotein expression in astrocytes following intracerebroventricular kainate injection. *Exp Brain Res* 1999;126:509–16.

[127] Seegers U, Potschka H, Loscher W. Expression of the multidrug transporter P-glycoprotein in brain capillary endothelial cells and brain parenchyma of amygdala-kindled rats. *Epilepsia* 2002;43:675–84.

[128] Lazarowski A, Ramos AJ, García-Rivello H, Brusco A, Girardi E. Neuronal and glial expression of the multidrug resistance gene product in an experimental epilepsy model. *Cell Mol Neurobiol* 2004;24(1):77–85.

[129] Kwan P, Still G, Butler E, Gant T, Meldrum B, Brodie M. Regional expression of multidrug resistance gene in genetically epilepsy-prone rat brain after single audiogenic seizure. *Epilepsy* 2002;43:1318–23.

[130] Girardi E, González NN, Lazarowski A. Refractory phenotype reversion by nimodipine administration in a model of epilepsy resistant to phenytoin (PHT) treatment. *Epilepsia* 2005;46 (Suppl. 6); 212:p606.

[131] Höcht C, Lazarowski A, Gonzalez NN, Auzmendi J, Opezzo JA, Bramuglia GF, Taira CA, Girardi E. Nimodipine restores the altered hippocampal phenytoin pharmacokinetics in a refractory epileptic model. *Neurosci Lett* 2007;413:168–72.

[132] Lazarowski A, Czornyj L, Lubienieki F, Girardi E, Vazquez S, D'Giano C. ABC transporters during epilepsy and mechanisms underlying multidrug resistance in refractory epilepsy. *Epilepsia* 2007;48(Suppl 5):140–9.

[133] Bauer B, Hartz AM, Pekcec A, Toellner K, Miller DS, Potschka H. Seizure-induced up-regulation of p-glycoprotein at the blood–brain barrier through glutamate and cyclooxygenase-2 signaling. *Mol Pharmacol* 2008;73:1444–53.

[134] Robey RW, Lazarowski A, Bates SE. P-glycoprotein–a clinical target in drug-refractory epilepsy? *Mol Pharmacol* 2008;73:1343–6.

[135] Hughes JR. One of the hottest topics in epileptology: ABC proteins. Their inhibition may be the future for patients with intractable seizures. *Neurol Res* 2008;30:920–5.

[136] Potschka H. Modulating P-glycoprotein regulation: future perspectives for pharmacoresistant epilepsies? *Epilepsia* 2010;51:1333–47.

[137] Tate SK, Depondt C, Sisodiya SM, Cavalleri GL, Schorge S, Soranzo N, Thom M, Sen A, Shorvon SD, Sander JW, Wood NW, Goldstein DB. Genetic predictors of the maximum doses patients receive during clinical use of the anti-epileptic drugs carbamazepine and phenytoin. *Proc Natl Acad Sci USA* 2005;102:5507–12.

[138] Gombert-Handoko KB. Treatment failure in patients with epilepsy—exploring causes of ineffectiveness and adverse effects. Thesis Utrecht University (with references), with summary in Dutch. ISBN/EAN: 978-90-393-5006-5.

[139] Su H, Sochivko D, Becker A, Chen J, Jiang Y, Yaari Y, Beck H. Upregulation of a T-type Ca^{2+} channel causes a long-lasting modification of neuronal firing mode after status epilepticus. *J Neurosci* 2002;22:3645–55.

[140] Huguenard JR. Block of T-type calcium channels is an important action of succinimide antiabsence drugs. *Epilepsy Curr* 2002;2:49–52.

[141] Aronica E, Yankaya B, Troost D, van Vliet EA, Lopes da Silva FH, Gorter JA. Induction of neonatal sodium channel II and III alpha-isoform mRNAs in neurons and microglia after status epilepticus in the rat hippocampus. *Eur J Neurosci* 2001;13:1261–6.

[142] Kuo CC. A common anticonvulsant binding site for phenytoin, carbamazepine, and lamotrigine in neuronal Na+ channels. *Mol Pharmacol* 1998;54:712–21.

[143] Remy S, Urban BW, Elger CE, Beck H. Anticonvulsant pharmacology of voltage-gated Na^+ channels in hippocampal neurons of control and chronically epileptic rats. *Eur J Neurosci* 2003;17:2648–58.

[144] Lucas PT, Meadows LS, Nicholls J, Ragsdale DS. An epilepsy mutation in the beta1 subunit of the voltage-gated sodium channel results in reduced channel sensitivity to phenytoin. *Epilepsy Res* 2005;64:77–84.

[145] Remy S, Gabriel S, Urban BW, Dietrich D, Lehmann TN, Elger CE, Heinemann U, Beck H. A novel mechanism underlying drug-resistance in chronic epilepsy. *Ann Neurol* 2003;53:469–79.

[146] Wadkins RM, Roepe PD. Biophysical aspect of P-glycoprotein mediated multidrug resistance. *Int Rev Cytol* 1997;171:121–65.

[147] Auzmendi J, Orozco-Suárez S, González-Trujano E, Rocha-Arrieta L, Lazarowski1 A. P-glycoprotein (P-gp) contributes to depolarization of plasmatic membranes of hippocampal cells in a model of phenytoin-refractory seizures induced by pentyleneterazole (PTZ). V° Latin-American Congress of Epilepsy (ILAE), Montevideo (ROU), November 5–8, 2008.

[148]   Bate L, Gardiner M. Molecular genetics of human epilepsies. *Exp Rev Mol Med* 1999;1: 1–22.

[149]   Steinlein OK. Genetic mechanisms that underlie epilepsy. *Nature Reviews Neuroscience* 2004;5:400–408.

[150]   Sillanpää M, Schmidt D. Natural history of treated childhood-onset epilepsy: prospective, long-term population-based study. *Brain* 2006;129:617–24.

Author Biographies

Liliana Czornyj, MD, completed her postgraduate training in clinical pediatrics and child neurology. She is a past president of SANI (Child Neurology Society of Argentina) and a member of the International Child Neurology Association (ICNA) and LACE (Epilepsy League of Argentina). For 25 years, she has worked at the Neurology Department of the Prof. Dr. Juan P. Garrahan Hospital Nacional de Pediatría of Buenos Aires, Argentina. She is coauthor of several textbooks of clinical child neurology, neuroinfectology, and neuropharmacology.

Alberto Lazarowski, PhD, is a biochemist trained at the School of Pharmacy and Biochemistry (FFyB) at the University of Buenos Aires, Argentina. He earned a masters in molecular biology and genetic engineering and a doctorate studying the role of folic acid in epileptic seizures in children. Currently, he is a professor of clinical biochemistry in the Department of Pharmacy and Biochemistry at the University of Buenos Aires. He was a postdoctoral fellow of the UICC (International Union Against Cancer), studying the MDR gene at UNC–Chapel Hill, and later he described for the first time the role of the *MDR-1* gene in Argentine children with refractory epilepsy. Later, he developed the confirmatory experimental studies in rats at the IBCN (Instituto de Biología Celular y Neurociencias, "Prof. E. de Robertis," School of Medicine, University of Buenos Aires).

Together, Dr. Czornyj and Dr. Lazarowski have studied several cases of pharmacoresistant epilepsies in children with overexpression of different ABC transporters in the brain epileptogenic areas associated with pharmacokinetic modifications of some AEDs and described the potential role of multidrug resistant proteins in refractory epilepsy and antiepileptic drug interactions. Both are members of GENIAR, an Ibero-Latin-American network of researchers in neuroscience (CYTED #610RT0405).